HELLO WEB DESIGN

by Tracy Osborn

**no starch
press**

San Francisco

Printed in the United States of America

First printing

24 23 22 21   1 2 3 4 5 6 7 8 9

ISBN (print): 978-1-7185-0138-6
ISBN (ebook): 978-1-7185-0139-3

Publisher: William Pollock
Executive Editor: Barbara Yien
Production Editor: Michele Mangelli
Interior and Cover Design: Tracy Osborn
Copyeditor: Sally Peyrefitte
Compositor: Happenstance Type-O-Rama
Proofreader: Betsy Dietrich

For information on book distributors or translations, please contact No Starch Press, Inc. directly:

No Starch Press, Inc.
245 8th Street, San Francisco, CA 94103
phone: 1-415-863-9900; info@nostarch.com
www.nostarch.com

Library of Congress Control Number: 2020945051

For my third book, it seems kind of redundant
to thank my husband *again*, and yet here we are.
Without him, this book (or any of my others)
wouldn't exist. Andrey, thank you for being my
#1 supporter and best friend.

Tracy Osborn is a designer, developer, and entrepreneur living in Toronto, Canada. Building websites since she was twelve, she always felt an affinity to computers, the internet, and what they bring us.

Tracy graduated with a BFA in Art & Design with a concentration in Graphic Design from California Polytechnic State University, San Luis Obispo, and worked as a web designer for five years before teaching herself programming and launching her first start-up, WeddingLovely.

She also speaks regularly at technology conferences, including keynotes at O'Reilly's Fluent Conference 2016, EuroPython 2017, and DjangoCon US 2017.

BRIEF CONTENTS

CONTENTS IN DETAIL

ACKNOWLEDGMENTS

This book has been a few years in the making. Thank you to everyone who has listened to my presentations, given me feedback, supported my ideas, and helped me make this book a reality.

In particular, thank you to everyone who took a monetary risk with me and backed the *Hello Web Design* Kickstarter campaign. With over 700 backers, this was my biggest campaign yet, and it means so much to me that folks are willing to support me with their dollars and cheer me on throughout the whole book production process.

Thank you as well to No Starch Press for taking a risk on me and adding this book to their collection. *Hello Web Design* will have more reach than I originally ever thought possible through their support, not to mention the countless hours they worked with me to improve the book and its content. I'm truly grateful for their partnership.

FOREWORD

Whenever I dipped my toe in the waters of the semantic web, I noticed there were two fundamentally different approaches. One approach was driven by the philosophy that absolutely everything in the universe should be theoretically describable. The other approach was far more lax, concentrating only on the popular use cases —people, places, events—and that was pretty much it. These few common items, so the theory went, accounted for about 80 percent of actual usage in the real world. Trying to codify the remaining 20 percent would result in a disproportionate amount of effort.

I always liked that approach. I think it applies to a lot of endeavours. Coding, sketching, cooking—you can get up to speed on the bare essentials pretty quickly and then spend a lifetime attaining mastery. But we don't need to achieve mastery at every single thing we do. I'm quite happy to be just good enough at plenty of skills so that I can prioritize the things I really want to spend my time doing.

Perhaps web design isn't a priority for you. Perhaps you've decided to double-down on programming. That doesn't mean forgoing design completely. You can still design something pretty good . . . thanks to this book.

Tracy understands the fundamentals of web design so you don't have to. She spent years learning, absorbing, and designing, and now she has very kindly distilled down the 80 percent of that knowledge that's going to be the most useful to you.

Think of *Hello Web Design* as a book of cheat codes. It's short, to the point, and tells you everything you need to know to be a perfectly competent web designer.

Say hello to your little friend.

—Jeremy Keith
Author of Resilient Web Design *and cofounder of Clearleft*

INTRODUCTION

This is not your typical beginner design book.

Beginner books tend to assume you eventually want to become an expert. Programming books assume you want to get a job in programming or computer engineering; design books assume you want to become a full-fledged designer.

But what if you wanted to learn *just enough* design to enhance your existing career?

I believe in the power of improving your skill set beyond your current focus. My personal experiences don't fit into any single category—I have a degree in Art & Design, I develop websites and apps using Python, and I have founded and managed a start-up. Knowing just enough in many areas has been crucial for my career.

Just as designers benefit from learning a little bit about programming, programmers (and marketers, and product managers, salesfolk, and so on) can benefit from learning a little bit about design. Even if you don't specialize in design, you will need to design at some point—whether it's working on slide decks, creating interfaces for programs and projects, or building a personal website. If you work with designers, a bit of design experience will give you a better foundation for communicating with them and understanding their work.

The purpose of *Hello Web Design* isn't to make you a designer—it's here to make you feel more comfortable doing design.

This book focuses solely on the visual stuff: you'll find no HTML, CSS, or frontend development and code within. We're going to work on making your designs feel more beautiful and, almost more important, how to make them *work* better.

Design is about problem solving. Even in places where you feel like you're just making things "prettier" (like working on a slide deck), you're really working to make the information in your slide deck easier and more pleasurable to read and comprehend. In our web interfaces, we really want to make things feel more natural and easier to use—good design is crucial to capturing and keeping new users and improving how successful our designs are.

Throughout this book, we'll explore not only theory (and the reasons behind some of those principles) but also shortcuts and quick solutions to some common design problems. For example, I'll share a bit about color theory but also some resources that will help you choose great-looking color palettes without starting from scratch. That pattern will repeat throughout the book—lots of shortcuts and some insight into why the shortcuts work.

My hope is that you'll feel more confident in doing design by the end of the book and then jump into more traditional beginner books covering design theory and practices more thoroughly if you want to become a traditional designer. Or, you can just use this book to benefit your current career. In either case, you'll be a much stronger designer by the end of this little book.

Let's get started!
    —**Tracy**

1 | IF YOU READ ONLY ONE CHAPTER, MAKE IT THIS ONE

IF YOU CAME ACROSS THIS BOOK and had time to read only one chapter, this would be the one for you. Why? Because it shows you the fastest way for you to improve your designs.

How it works is more important than how it looks

I know you probably came to this book looking for help with how to improve the look of your designs, but stick with me for a moment: the way your design *works* is more important than how it *looks*.

Take Craigslist, for example. The website largely hasn't changed over the years, and the design appears very dated (**FIGURE 1-1**).

If how it looked were most important, Craigslist would have lost out years ago to any of the other shiny new classified ad websites that have popped up.

Craigslist has continued to be the top destination for classified ads simply because it's so simple and straightforward to use.

FIGURE 1-1: Craigslist is still the top of its class without a modern-looking website design.

With no flashy effects or distracting banners, Craigslist makes it super simple to post ads and search existing ads.

If you're fretting that your website is too simple and doesn't look "modern"—that is, doesn't adhere to all the latest trends, which don't necessarily dictate "good" website design—I'm here to reassure you that looks matter far less than whether your design succeeds in what you and your users want it to achieve.

There are two important steps for ensuring that your design results in a good user experience: knowing what you want your design to accomplish and getting feedback from others.

Determine what success looks like

You can't track how well your design is doing its job until you know your design's goal. Different kinds of designs have different goals. For example, you might want visitors to:

- Fill out a form.

- Spend an average of at least 30 seconds viewing a page.

- Subscribe to a newsletter.

- Leave a comment.

You might have heard the term *conversion rates* in user experience books, which is a fancy way of asking, "What percentage of people are doing what I want them to do?" Once you've defined what "success" looks like for your website, you can make better decisions about your design, measure how well your current design works, and improve your conversion rate.

Ask other people to view and review your designs

If you're feeling bashful about your newfound design skills, you might not want to show your designs to anyone else. What if they don't like them? What if they say something negative?

Getting over these feelings of insecurity and getting your designs in front of people who aren't you will lead to the *biggest* improvements in your designs. As a designer, you cannot objectively evaluate how well your designs work. You become blind to potential problems.

By showing your designs to others, you'll be able to see whether people successfully achieve what you want them to, and whether anything about the site is confusing. And you may get some high fives as well.

Feedback is useful, but don't feel you need to follow every suggestion or address every comment you receive. File them away in your mind, or keep a list, and then take care of all of those changes once you've heard from several people. Sometimes you'll get contradictory feedback from two different people, and that's fine (if one of them more closely fits the profile of your target user, that person's feedback is more important!). Maybe there's a third approach that could make everyone happy, or maybe it's not worth worrying about that piece yet.

Quick tips on improving your design

"But Tracy," you might be saying, "I really want to make my design *look* better too, not just work better."

I hear you! The way your design looks also affects user experience, and you can improve your design by keeping one key concept (which applies to many things) in mind: *cut down on clutter*.

Cut down on clutter for better looking designs

Clutter—such as you see in busy, disorganized, and chaotic designs—is the bane of good user experience, and cleaning up your clutter creates websites that both *look* better and *work* better.

Expanding on this concept, here are some short, easy-to-understand principles that we'll cover in more depth later:

USE A GRID

Most designs cut down on clutter by using a grid, which creates a skeleton for your design to anchor objects and create an unconscious feeling of order. Small pixel differences can make a design feel sloppy and unprofessional.

We'll cover this topic in more depth in Chapter 2, Section 2.1.

CHOOSE JUST A FEW COLORS

Designs that are chock-full of different colors (24 shades of blue, 5 shades of red, and so on) look busy. Set a specific color palette and use only those colors in your design to create a much more cohesive look.

We'll cover this topic in more depth in Chapter 2, Section 2.2.

LIMIT YOURSELF TO TWO DIFFERENT TYPEFACES

While you *can* use one typeface for your navigation, another for your text, another for your buttons, and another for your headlines—this will make your designs feel chaotic. Restrict the number of typefaces you use, and use bolding, italics, all caps, and other transforms to create variety and indicate emphasis.

We'll cover this topic in more depth in Chapter 2, Section 2.3.

SIMPLIFY YOUR TEXT

Large blocks of text come across as clutter. Simplify your sentences and limit paragraphs to two to three sentences. You can also break text up by adding bullets and headlines. Readers on the web tend to skim text, and shorter paragraphs keep more readers reading.

We'll cover this topic in more depth in Chapter 2, Section 2.6.

ADD WHITE SPACE

White space is the ultimate clutter reducer. One of the biggest mistakes new designers make is pushing items much too close to each other. Adding space between elements (content, widgets, forms, buttons, images, text, and so on) makes designs easier to read and more modern, airy, and inviting.

We'll cover this topic in more depth in Chapter 2, Section 2.4.

All of the above can be distilled into one sentence:

> **Make sure your design *works* well,
> and reduce as much clutter as possible.**

This advice puts you on the right path to creating better designs, and the upcoming chapters explore all of these guidelines in more detail.

2 | CHAPTER 2
THEORY AND DESIGN PRINCIPLES

LET'S DIVE INTO THIS INFORMATION IN MORE DEPTH, shall we?

Each section in this chapter is going to balance theory (so you understand the background and the why of things), real-life examples (so you can see these principles being applied), and shortcuts (so you can apply some of these principles quickly without starting from scratch).

I'll give you a few rules and suggestions to remember. Of course, there are exceptions to every rule, and every rule can be broken. Because these sections are aimed at beginners, we'll be sticking to these principles, but as you grow as a designer you'll learn when you can break some of the rules.

As we go through each section in depth, we'll apply these principles to the little widget (shown in FIGURE 2-1).

FIGURE 2-1: A (rather ugly) widget. Let's improve how it looks, shall we?

It's not much to look at right now, but it will gradually look better as we apply what we learn.

Let's get started!

GRID

OUR FIRST PRINCIPLE IS AN EASY ONE: line things up!

You may have heard this common designer complaint:

A designer builds a pixel-perfect mock-up and passes it along to the developer to build. The developer builds the design, but the built design differs from the mock-up by something arbitrary like two pixels!

"Silly designer," says the developer. "Why does something so small matter? It's basically the same."

Here's the thing—small pixel differences really do matter, especially when it comes to elements on a page. If one element is close to but not completely lined up with another element, it can create a tiny bit of unevenness, and that little bit of chaos leads to a feeling of unease and clutter (FIGURE 2-2).

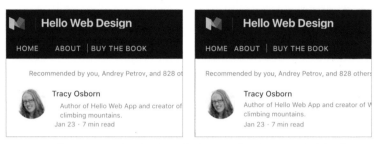

FIGURE 2-2: The two screenshots are basically the same, but the one on the left, where everything is slightly out of line, looks less cohesive and professional.

The easiest way to make sure that elements line up with one another is to add a grid to our design. A grid adds an invisible skeleton to our app, a scaffolding that we can use to set up and arrange our elements so the overall layout looks tidy and ordered. Aligning elements to an underlying grid will help you achieve alignment and consistency throughout your design.

The *New York Times* website **(FIGURE 2-3)** uses five main *columns* to organize a large amount of information. The spaces between the columns highlighted in red are *gutters*. Objects in your grid can span multiple columns, and some objects might break out of the grid, but everything more or less adheres to the grid.

You can use any number of columns in your design **(FIGURE 2-4)**, but a 12-column grid **(FIGURE 2-5)** is the most often used for versatility.

A grid also helps you create a plan for your website's layout because it constrains you to place elements within certain areas rather than having the entire page to choose from. Win win win.

In essence: Got a bunch of elements? Line them up (horizontally, vertically, or both) to create a feeling of cohesiveness.

FIGURE 2-3: The *New York Times* website uses a grid to effectively organize a large number of elements and showcase items on its home page.

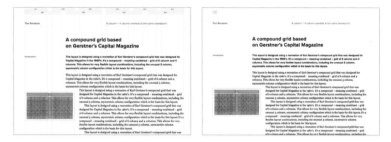

FIGURE 2-4: The Gridset website demos a compound 4+6 grid with and without guidelines.

FIGURE 2-5: The 12-column grid system in the Bootstrap CSS framework.

Shortcuts

The theory here is simple—just line things up—but there are many tools that make it easy to ensure we're using the grid.

Grids in mock-up programs

If you're working on something that you won't be building in CSS, you'll need to add guides to your design.

All mock-up programs—Photoshop, Sketch, or GIMP— allow you to set guides to float over your design, which makes it easy to align elements to the guide.

If you're working on a website mock-up in something like Photoshop, you can use grid templates to lay out your website using the same columns you'd use in your website framework system (FIGURE 2-6).

FIGURE 2-6: Guides can be set by dragging from the rulers at the left and top of the screen in Photoshop.

Guides are also accessible in most slide programs and other simple layouts, such as Keynote (**FIGURE 2-7**).

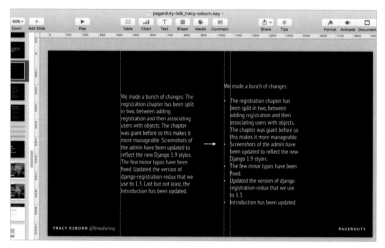

FIGURE 2-7: Keynote also lets you pull out guides from the rulers on the screens. Many other programs have similar options.

For slides, you don't need to add a whole 12-column grid to your system—the slides in FIGURE 2-7 have only a few guides, just enough to keep elements on separate pages along the same lines.

However, if you're designing something a bit more complex, you can download quite a few templates with multiple columns already set up, such as the 960.gs (*hellobks.com/hwd/4*) system.

Grids for web design

I highly recommend using a CSS framework with a grid included, such as Bootstrap (*hellobks.com/hwd/5*), Foundation (*hellobks.com/hwd/6*), Skeleton (*hellobks.com/hwd/7*), mini.css (*hellobks.com/hwd/8*), or PureCSS (*hellobks.com/hwd/9*) (**FIGURE 2-8**).

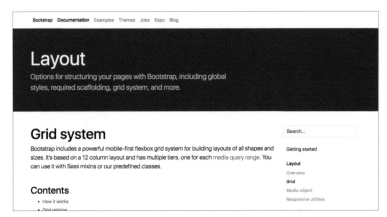

FIGURE 2-8: The grid system included in the Bootstrap CSS framework.

By using the HTML classes that constrain your design to the underlying CSS Grid, your elements will naturally align with other elements on the page **(FIGURE 2-9)**. Just keep in mind that additional margins or padding that might be added in CSS may bump elements out of alignment.

.col .col-md-8			.col-6 .col-md-4
.col-6 .col-md-4	.col-6 .col-md-4		.col-6 .col-md-4
.col-6		.col-6	

FIGURE 2-9: Some of the CSS classes included in Bootstrap to align and set your objects in the underlying grid.

CSS is getting a new element called CSS Grid (how convenient!) that makes it super simple to align elements to a grid without using a CSS framework. At the time of this book's writing, CSS Grid is on the verge of being released and covered by most browsers. While this book doesn't cover CSS, using CSS Grid will make it easier to implement grid-based designs **(FIGURE 2-10)**.

FIGURE 2-10: CSS Grid docs on the Mozilla Developer Network (*hellobks.com/hwd/10*)

Real-life examples

Remember the little widget we introduced at the beginning of this chapter? Let's update the design by lining up the elements **(FIGURE 2-11)**.

FIGURE 2-11: On the left, the original widget. On the right, we've lined up the edges of the elements in the box, making it feel a bit cleaner and less chaotic.

While the widget still has issues (after all, we're just getting started), you can see that this little change has made a small but significant positive difference. All of the interior elements are now lined up: headline, content, input, button. The input's placeholder text is bumped out a bit because of the padding in the form, but its containing element aligns to the grid. The overall feeling is just a bit more cohesive, a little less chaotic.

Again, in a nutshell: just line things up. Add guides to align your elements to an invisible grid, and pay attention to the little pixel differences and misalignments that can make a design feel more chaotic. Feel free to break the rules occasionally and go off the grid (*ha*), but using a grid for most of your elements will create a cleaner, more organized design.

Next up, we're going to talk color!

COLOR

COLOR IS SO IMPORTANT TO DESIGN, but it's a topic that can be overwhelming. With all of the rainbow at our fingertips, how do we choose the right colors? There's a reason why color theory is a semester-long course for design students.

In my own color theory course, I had to create a perfect gradient in 20 swatches from black to white in paint. The best way to accomplish that was to paint 200+ swatches: taking black paint, add a drop of white, paint a swatch . . . add another drop of white paint, paint another swatch . . . and then narrow down those hundreds of swatches into 20 perfect, even steps from white to black.

Don't worry, I'm not going to put you through that. Look how much time and effort you're saving by reading this book!

Color theory courses cover a lot of terms that we're going to skip here—CMYK vs. RGB colors, color harmony theory and color wheels, triadic and analogous colors—it can get really overwhelming really fast. I'm going to cover a very broad overview of color theory aimed at getting you more comfortable as

RED: aggressive, important, passionate

ORANGE: energetic, playful, affordable

YELLOW: friendly, happy, attentive

GREEN: growing, natural, successful

BLUE: trustworthy, comforting, relaxed

PURPLE: luxurious, romantic, mysterious

PINK: playful, innocent, youthful

BROWN: stable, rustic, earthy

BLACK: powerful, sophisticated, edgy

WHITE: virtuous, sterile, healthy

GRAY: formal, neutral, professional

IVORY: quiet, calm, elegant

fast as possible; at the end of the chapter I'll give you resources that you can use to learn more about everything we cover.

First up is color psychology, a color "hack" that you can use to invoke various feelings in your work. Color can affect how your design is felt and perceived. In a very general sense, these are the emotions evoked by certain colors:

- **Red:** aggression, importance, passion
- **Orange:** energy, playfulness, affordability
- **Yellow:** friendliness, happiness, attention
- **Green:** growth, nature, success
- **Blue:** trust, comfort, relaxation
- **Purple:** luxury, romance, mystery

- **Pink:** playfulness, innocence, youth

- **Brown:** stability, rusticity, earthiness

- **Black:** power, sophistication, edginess

- **White:** virtue, sterility, health

- **Gray:** formality, neutrality, professionalism

- **Ivory:** quiet, calm, elegance

When you want your design to evoke a feeling of success and stability, green and blue are great colors to use. A website for a sleek modern hotel might use black and purple. In general, warmer colors (reds, yellows) are more energetic and exciting, whereas cooler colors (blues, purples) are more stable and calming.

It's worth noting that these are Western associations—if you're designing for another culture, it would be worth researching any color connotations that specific culture may have. For example, white is a mourning color in China, compared to black in Western cultures.

You can also play with vibrancy. Less vibrant hues, like those on the "Keep Portland Weird" page on the Keep Earthquakes Weird website by Oblio, are calmer and more subtle than bright and vibrant hues, like those used on the Citysets website. In color psychology, blues are relaxing, but the bright, eye-popping blue feels more energetic (FIGURE 2-12).

These are not hard-and-fast rules. You can certainly create a calm and elegant website using red—however, these color psychology shortcuts give you a jumping off point so you're not overwhelmed with color options at the start.

FIGURE 2-12: Less vibrant hues are calmer and less energetic than bright, eye-popping versions.

Shortcuts

Now that we know some fundamentals, let's check out a few shortcuts we can use to save ourselves time.

Limit your color scheme

Avoid picking colors at random for your design. Choosing two to four colors for your site and limiting all of your elements to the colors on your palette will help your site feel less busy **(FIGURE 2-13)**.

Use color to make your design "pop"

When building a color palette, avoid the temptation to go for all bright colors. Using mostly grayer or more neutral colors with one "pop" color makes it easier to highlight important elements without creating a chaotic design **(FIGURE 2-14)**.

Pay attention to contrast

Light gray text on a white background looks pretty but is a nightmare to read, especially for readers who may have a visual impairment **(FIGURE 2-15)**. Colored text on a colored background can also get you in trouble if there isn't enough contrast.

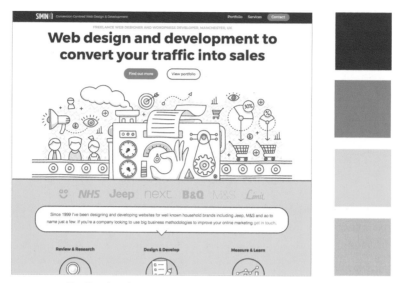

FIGURE 2-13: The Siminki website has a limited color scheme that helps it feel more unified.

FIGURE 2-14: The Habita website has a strong pop of color.

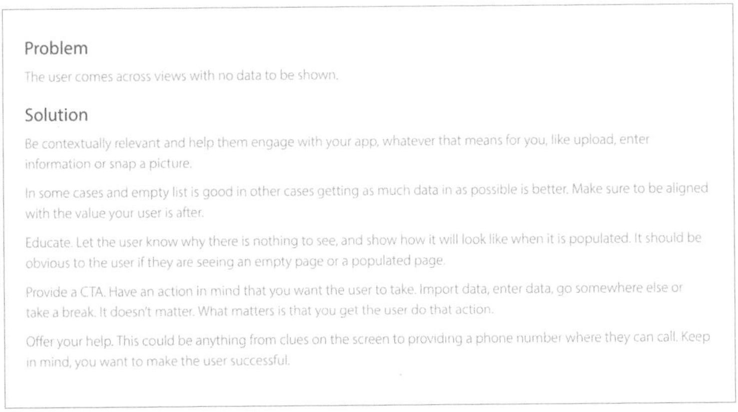

FIGURE 2-15: Light gray text on a white background has low contrast and can be hard to read.

When in doubt, use a color contrast checker such as WebAIM (*hellobks.com/hwd/25*) to make sure your text is readable.

Use a color palette website

Building a color palette from scratch can take a long time and a lot of thought. Thankfully, a ton of websites are dedicated to helping you choose a color palette to use in your designs.

Adobe Color CC (*hellobks.com/hwd/28*) builds a palette of colors around your choice of a base color and various color schemes—I recommend "complementary" (**FIGURE 2-16**).

The Material Design Palette website (*hellobks.com/hwd/29*) automatically displays colors in a sample design, which is super helpful when you're struggling to visualize how two colors could work together (**FIGURE 2-17**).

A new option is Colormind (*hellobks.com/hwd/30*), which uses deep learning to generate color palettes. You can loop through palettes as well as generate palettes after setting one or more specific colors (**FIGURE 2-18**).

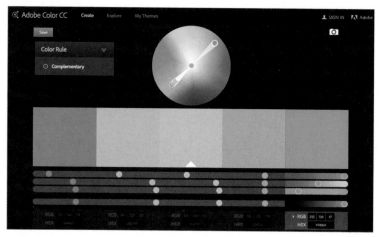

FIGURE 2-16: Adobe Color CC (*hellobks.com/hwd/28*)

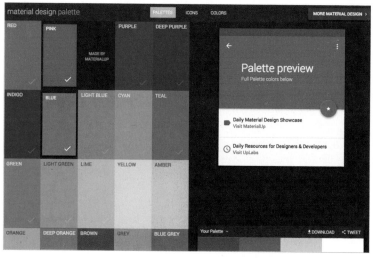

FIGURE 2-17: Material Design Palette website (*hellobks.com/hwd/29*)

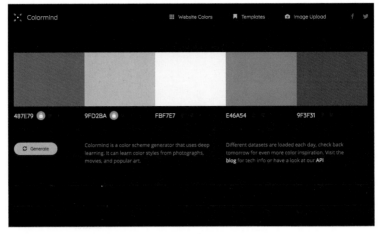

FIGURE 2-18: Colormind (*hellobks.com/hwd/30*)

Keep in mind that you can alter the colors in a preset palette slightly if you wish—they aren't set in stone. I often find that I need to lighten or darken a shade in order to get good contrast in my design.

Real-life examples

Don't get too married to a particular color palette—once you implement the colors into your design, you might find that your original favorite doesn't work as well as you had hoped. I usually go through at least a few palettes until I find one that feels best.

Let's run our little widget through some different color palette options **(FIGURE 2-19)**.

That last option gives our text good contrast, a nice color "pop" color for our button, and a soothing green background—looking pretty good, and certainly better than the original version with no color at all!

FIGURE 2-19: Applying different ideas and options until we find something that feels right.

Want to read more? This article from *Smashing Magazine*, "A Web Developers Guide to Color," is my favorite article on color theory for web: *hellobks.com/hwd/31.*

With millions of colors, it can feel really hard to narrow your options down to a nice palette for your website. Make it easier for yourself by using a color palette generated by the options we mentioned, which will guarantee you a harmonious color palette to use in your designs.

Next up, another big subject—typography and text!

TYPOGRAPHY

IF YOUR DESIGN INCLUDES ANY KIND OF TEXT, you're dealing with typography. According to Wikipedia, "Typography is the art and technique of arranging type to make written language legible, readable, and appealing when displayed." Simply put, make your words easy and enjoyable to read.

Type can be beautiful but hard to read (websites with trendy thin, pale gray fonts) but also easy to read and less pleasing to look at (pretty much any website using default system fonts). Succeeding in both areas can seem complicated, but there are a lot of great shortcuts that make it easier.

We could spend another whole university semester covering typography theory in full, the history of typefaces, and more. Instead, I'm going to pack a lot of this information into one quick section.

You might be wondering about the difference between a *font* and a *typeface*. To be completely correct, a typeface is a "family," or collection of fonts. *Arial* is a typeface, and *Arial Bold* is a font. That said, the two terms are frequently used interchangeably (unless you're a typographer), and we'll be doing so here.

Typography fundamentals

Let's start by covering some important terms and concepts about typography.

Typeface categories

The main difference between fonts is whether or not they include *serifs*, which are the little decorative bits at the end of letters. This book's content text uses a serif font (Tisa Pro), while the headlines use a sans-serif font (literally *without* serif—and Tisa Sans Pro is the headline font here) **(FIGURE 2-20)**.

serif

FIGURE 2-20: Sans-serif versus serif typefaces. Tisa Sans Pro on the left, Tisa Pro on the right.

In general, serif fonts are easier to read in printed text, and sans-serif fonts are easier to read on a screen. The serifs in printed text help the eye quickly read the letters, whereas on screen it's harder to render the tiny serifs, which end up hindering readability (though, with the advent of newer and crisper screens, this distinction is going away).

Fonts can be further broken down into several categories **(FIGURE 2-21):**

- **Slab serif:** Serifs are "slab"-like—more square and angular than typical serifs.

- **Monospace:** Every letter takes up the same amount of space. In other typefaces, thin letters such as "i" take up less space than wide letters such as "m."

- **Display:** Fancy, often swooshy fonts, not particularly readable but nice when used sparingly to create emphasis in headlines or large text.

- **Handwriting**: Fonts that look like someone's handwriting. These are often lumped together with display fonts.

Ab Ab **Ab** Ab

FIGURE 2-21: Slab serif (Chaparral Pro), monospace (Courier New), display (Buttermilk), and handwriting (HanziPen TC) typefaces.

Leading and line height

Make sure your lines of text aren't too close together, which creates a feeling of clutter. The space between lines of text in a paragraph is known in typography-land as "leading" (from the strips of lead that typographers added between the lines of text when they hand-set type into the printing press). In CSS the same concept is referred to as "line height" **(FIGURE 2-22).**

Aim for a Goldilocks amount of spacing—a paragraph is hard to read if the lines of text are too close or too far apart. Give lines of text a little space to improve readability and reduce the feeling of clutter, but not so much that they become hard to read. 1.6 is a good starting ratio (e.g., 12px font size with 19.2px line height, 14px font size with 22.4px line height). Feel free to

Make sure your lines of text aren't too close together, which creates a feeling of clutter. The space between lines of text in a paragraph is known in typography-land as "leading"

Make sure your lines of text aren't too close together, which creates a feeling of clutter. The space between lines of text in a paragraph is known in typography-land as "leading"

Make sure your lines of text aren't too close together, which creates a feeling of clutter. The space between lines of text in a paragraph is known in typography-land as "leading"

FIGURE 2-22: Text can be too close or too far apart, hindering readability. The center example is the most readable.

play around with your spacing to find what feels right for your specific situation.

Kerning and letterspacing

Kerning is the process of changing the spacing between individual letters, and letterspacing is the spacing between all letters (changing it affects every letter, whereas kerning affects only a pair of letters). As with leading, we don't want too much or too little. Thankfully, the defaults set for us when text is displayed in browsers are fairly ideal, but CSS allows you to adjust the amount of spacing between letters using the letterspacing property. Graphics programs also allow you to tinker with kerning. Large headlines sometimes look nicer with some space removed, whereas small text can benefit from adding some space (**FIGURE 2-23**).

Kerning is the process of changing the spacing between individual letters, and letter-spacing is the spacing between all letters (changing it affects every letter, whereas kerning affects only a pair of letters).

Kerning is the process of changing the spacing between individual letters, and letter-spacing is the spacing between all letters (changing it affects every letter, whereas kerning affects only a pair of letters).

FIGURE 2-23: Letters that are too close to one another make words and sentences harder to read.

Typography principles

Now, let's cover some important principles to keep in mind as you work on typography.

Limit designs to two typefaces

Using lots of typefaces can lend to chaos and clutter. Simplify by choosing one typeface for headlines and a second for content; this will make your design look much cleaner. You can use bolding, italics, uppercase, and other stylistic treatments to create more styles in your text (FIGURE 2-24).

FIGURE 2-24: Lots of typefaces make a design feel chaotic (left). Fewer typefaces are easier to wrangle and look more professional.

Avoid justifying or centering text

Justified text is text that fits entirely in a column; both margins are justified rather than just one. Based on Section 2.1, where we discussed lining things up, you might be tempted to use justified text to create an even line on the right of the text (FIGURE 2-25).

However, justified text creates some serious issues:

- Spacing words out to fit a column can add large, unsightly spaces between words.

- Large spaces between words can line up across multiple lines of text, creating an unsightly visual "river" of white space (FIGURE 2-26).

Justified text is text that fits entirely in a column; both margins are justified rather than just one. Based on Section 2.1, where we discussed lining things up, you might be tempted to use justified text to create an even line on the right of the text

Justified text is text that fits entirely in a column; both margins are justified rather than just one. Based on Section 2.1, where we discussed lining things up, you might be tempted to use justified text to create an even line on the right of the text

FIGURE 2-25: "Ragged-right" text (right), is easier to read than justified text (left).

It was the White Rabbit, trotting slowly back again, and looking anxiously aboutas it went, as if it had lost something; and she heard it muttering to itself `The Duchess! The Duchess! Oh my dear paws! Oh my fur and whiskers! She'll get me executed, as sure as ferrets are ferrets! Where CAN I have dropped them, I wonder?' Alice guessed in a moment that it was looking for the fan and the pair of white kid gloves, and she very good-naturedly began hunting about for them, but they were nowhere to be seen--

It was the White Rabbit, trotting slowly back again, and looking anxiously about as it went, as if it had lost something; and she heard it muttering to itself `The Duchess! The Duchess! Oh my dear paws! Oh my fur and whiskers! She'll get me executed, as sure as ferrets are ferrets! Where CAN I have dropped them, I wonder?' Alice guessed in a moment that it was looking for the fan and the pair of white kid gloves, and she very good-naturedly began hunting about for them, but they were nowhere to be

FIGURE 2-26: Justified text without hyphenation, on the left, has unsightly gaps and rivers of white space (marked in red).

Unless you have a lot of time for tweaking, justified text is not worth it. Use left-aligned (otherwise known as ragged-right) text to ensure easy readability.

Centered text can be challenging

Centered text can work for headlines, but when in doubt, left align your text against an underlying grid so it all lines up for easier readability. Paragraphs of text are *especially* hard to read when centered because the starting point for each line changes (FIGURE 2-27).

Centered text can work for
headlines, but when in doubt, left
align your text against an
underlying grid so it all lines up for
easier readability.

FIGURE 2-27: The uneven left edge of centered content makes it harder to read.

Line length

A paragraph becomes significantly harder to read if there are more than 75 or fewer than 45 characters per sentence. Make sure your paragraphs have an appropriate width (or font size) so that the lines are just long enough for maximum readability (FIGURE 2-28).

It was the White Rabbit, trotting slowly back again, and looking anxiously about as it went, as if it had lost something; and she heard it muttering to itself `The Duchess! The Duchess! Oh my dear paws! Oh my fur and whiskers! She'll get me executed, as sure as ferrets are ferrets! Where CAN I have dropped them, I wonder?'

It was the White Rabbit, trotting slowly back again, and looking anxiously about as it went, as if it had lost something; and she heard it muttering to itself `The Duchess! The Duchess! Oh my dear paws! Oh my fur and whiskers! She'll get me executed, as sure as ferrets are ferrets! Where CAN I have dropped them, I wonder?'

FIGURE 2-28: Overly long lines in paragraphs are hard to read.

Shortcuts

Now that you've absorbed this new information, let's cover some shortcuts that'll make designing your typography much faster.

Free font resources

The most beautiful fonts usually cost a lot of money—worth it to professional designers but perhaps not to hobbyist designers like ourselves.

Thankfully, sites such as Google Fonts (*hellobks.com/hwd/32*) **(FIGURE 2-29)** and Adobe Fonts (*hellobks.com/hwd/33*) **(FIG-URE 2-30)** offer us beautiful font options that we can use in both online and print designs. Google Fonts allows you to download the fonts, and Adobe's Creative Cloud feature makes Adobe Fonts' web fonts available in design and slide programs.

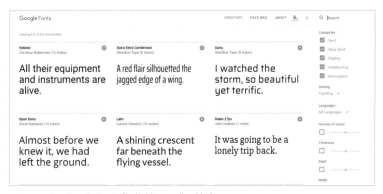

FIGURE 2-29: Google Fonts (*hellobks.com/hwd/32*)

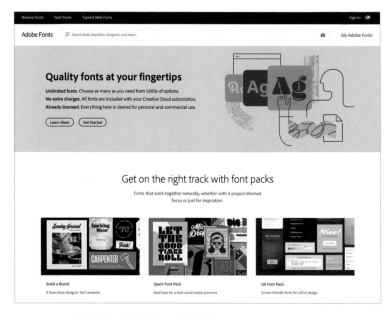

FIGURE 2-30: Adobe Fonts (*hellobks.com/hwd/33*)

Adobe product screenshot(s) reprinted with permission from Adobe Systems Incorporated.

Portions of this page are reproduced from work created and shared by Google and used according to terms described in the Creative Commons 3.0 Attribution License.

Other free font websites exist, but these two consistently have the largest collection of fonts that look great and are easy to use.

Curated font websites

Thousands of fonts and typefaces exist in Google Fonts and Adobe Fonts alone—how do you choose the best options for your design?

Rather than scrolling through hundreds of options and reading the same default sentence over and over (guaranteed to make anyone's eyes cross), check out sites where designers have curated great-looking free fonts to narrow your choices.

These websites make it easier to choose typefaces that work for your design and help you choose fonts that are beautiful and professionally designed (FIGURE 2-31).

Beautiful Web Type (hellobks.com/hwd/34)

Typewolf (hellobks.com/hwd/35)

Brick.im (hellobks.com/hwd/36)

Font Pair (hellobks.com/hwd/37)

FIGURE 2-31: Various websites that curate free fonts and recommend best options.

Real-life examples

Let's grab that widget and limit ourselves to just two typefaces: one serif and one sans serif (FIGURE 2-32).

FIGURE 2-32: New, cohesive fonts make our widget look more professional.

The new fonts (Tisa Pro and Tisa Sans Pro) look more professional and play well together than the four separate fonts that the widget used before.

Like color, typography is a huge, fascinating subject. Hopefully this brief section has given you a good overview that gets you started and gives you a good platform for learning more.

By narrowing down your font choices to those that have already been vetted by others and reducing the number of fonts in your design, you'll be well on your way to a design with beautiful typography.

Next up, we're going to cover white space and giving your design room to breathe.

WHITE SPACE

IF YOU USE ONLY ONE TOOL TO IMPROVE YOUR DESIGN, I guarantee that white space would make the biggest difference. White space is the *ultimate* clutter reducer.

Also known as negative space, white space is empty space (not necessarily white). Essentially, white space is the space in your page left blank as well as the space between your elements (FIGURES 2-33, 2-34, AND 2-35).

When you're designing, you might feel tempted to fill up your design with information, links, and other useful elements. After all, empty space can feel like wasted space, a place where you could add more content to convince your readers to stay longer, use your website, or buy your product. Why not fill up that space?

However, a cluttered, crowded website—even one with more information—will perform worse than a website that makes up in simplicity and breathing room what it loses in information. White space is crucial to improving how your design is perceived and how well it works.

FIGURE 2-33: Let's inspect the white space on the Squarespace website.

FIGURE 2-34: White space, highlighted in red, is the space on the page without content. Background images and textures generally count as white space.

FIGURE 2-35: White space also refers to space between elements and lines of text.

White space fundamentals

Here are a few more reasons why white space is important.

White space improves comprehension

Cluttered websites and designs are harder to follow and overload the user with too much information. Simpler websites make it easier for users to digest the content you have displayed without being overwhelmed by extra, unnecessary detail (FIGURE 2-36).

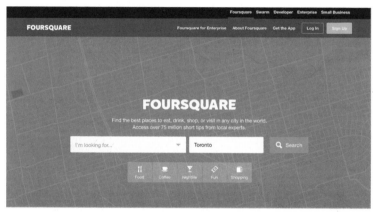

FIGURE 2-36: The Foursquare website has plenty of white space, letting the user know what to read and what to do next. Simplicity makes designs less confusing.

White space improves legibility

We're going to repeat a bit of information we covered in the last section just to make sure it sticks.

Remember when we covered leading (the space between lines) in the typography section? White space in the form of leading improves readability. Too little mushes sentences together, but too much leading can also worsen readability—this is definitely a case where you should shoot for the Goldilocks zone. Adding space between lines of text and elements will help users read what you're trying to say (FIGURE 2-37).

Too little space between lines in a paragraph (leading) makes the paragraph hard to read quickly.

Give your lines a bit of breathing room (as mentioned in the typography section) for better readability!

FIGURE 2-37: Paragraphs with too little space between lines are hard to read.

Similarly, the space between letters (kerning) also matters. Making sure your letters aren't too squished together and have enough space between them will make your text easier to read (FIGURE 2-38).

Not enough space between letters also hinders readability.

Make sure you keep the natural spacing between letters!

FIGURE 2-38: Letters that are too close together also are hard to read. Add some space to improve readability.

White space improves calls to action

More items crowding a space make it a harder for someone to identify your call to action (also known as your CTA). Whatever element you want your reader to use, they'll be more likely to see and use if it stands out with a lot of white space—drawing attention to the CTA while reducing distractions from other elements (FIGURE 2-39).

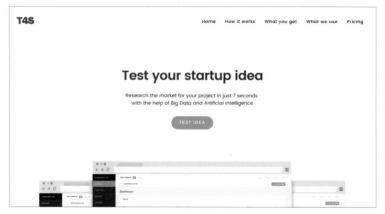

FIGURE 2-39: The introduction text and button stand out with copious amounts of white space in the T4S website.

White space helps set a tone

Open, blank space is associated with luxury and quality, whereas designs with less space and elements pushed up against each other is associated with a more thrifty, chaotic tone—like a busy flea market versus a luxury brand storefront.

If you want your design to give the impression of luxury, professionalism, elegance, or superiority, you can bolster that impression by using white space liberally **(FIGURES 2-40 AND 2-41).**

FIGURE 2-40: Cluttered websites lacking whitespace feel cheap.

FIGURE 2-41: The Helm website for yachting has tons of open space and feels luxurious.

Quick theory

As a new designer, your instinct is going to push you toward less space, not more. Let's look at where you can add white space in your designs.

Space between lines of text

White space starts at the content level. Your text needs space to breathe. Make sure you give the lines of text enough room to improve readability (FIGURE 2-42).

FIGURE 2-42: White space in the form of space between lines of text.

Space between elements

Now that we have our text figured out, let's focus on the spacing in between elements. This means the space between your content and paragraphs and other paragraphs, or paragraphs and their headline—any element near another element (FIGURE 2-43).

Spacing between individual elements helps set the elements apart from each other, so the eye can more effectively pick out the separate groups. In FIGURE 2-43 the space between the

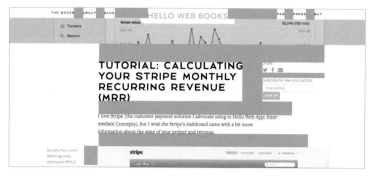

FIGURE 2-43: White space in the form of space between elements.

newsletter signup (to the right of the article title) and the title ensures that the two elements remain distinct objects.

Space between groups of elements

Taking the macro view, we also want to make sure *groups* of elements have enough space around them. This means creating space between columns of elements and other columns, rows and other rows, and your entire page of elements within its container (such as the browser window) **(FIGURE 2-44)**.

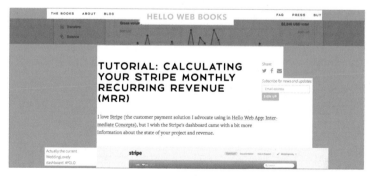

FIGURE 2-44: White space in the form of space between groups of elements.

Shortcuts

Here are a couple of easy tips to help you use white space more effectively in your designs.

Double white space between elements

Use more white space than you think you need! Yes, there can be such thing as too much, but using more than you think is a good rule of thumb when you're starting out. Try to double what your initial instinct tells you is appropriate. Go farther than you're comfortable, take a break, and then re-evaluate. What might have felt too airy at first may now feel natural and balanced.

Make sure you have enough leading in your paragraphs

Repeating our typography section again for emphasis! Remember to keep enough leading (vertical white space between lines in a paragraph) so your paragraph is easy to read.

Real-life examples

Removing white space from my personal website illustrates the how drastically white space can alter a design. Without white space, the site looks cluttered and amateur. Adding white space, especially in the very airy middle, makes the design look more professional and deliberate (FIGURE 2-45).

Let's add more white space to our widget: between individual lines of text and between individual elements within the widget (FIGURE 2-46).

Doesn't that feel a lot cleaner? I also evened out the spacing between elements (the same amount of space around the elements within the container, as well as the same amount of

FIGURE 2-45: Before and after adding white space to my personal website. An airy design feels less chaotic.

FIGURE 2-46: Adding whitespace between lines of text, between elements, and around elements helps our widget feel more professional.

space between the headline and the content, and the content and the form). This even amount of spacing also reduces the clutter and chaos.

Type, color, and lining things up can do a lot, but adding breathing room is the biggest tool in your decluttering basket to build a better-looking design.

Next up, we're going to tackle layout as a whole—moving from these micro issues and starting to take a macro view of our design!

LAYOUT AND HIERARCHY

THE LAST FEW SECTIONS DEALT WITH CONCRETE CONCEPTS: color, type, and white space. Now we're going to consider more abstract concepts: layout and hierarchy.

Layout refers to the way that information is arranged and prioritized; the choices that you make about layout affect how well your message is read and understood. We already covered the grid: lining up elements to reduce the little pixel differences that can make a design feel cluttered. Positioning and hierarchy of elements are other important aspects of layout.

Elements could go literally anywhere—how do you identify their proper places?

As English readers, we read left to right, meaning that our eyes naturally start at the top left of a page. Studies have shown that website users look at websites in an "F" pattern **(FIGURES 2-47 AND 2-48)**.

Take advantage of this discovery and build layouts that start at the top left of the page and continue to the right and down

FIGURE 2-47: The Google Ventures website is a good example of the "F" pattern.

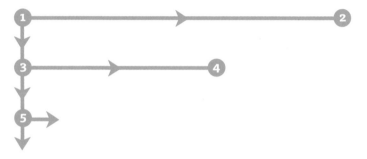

FIGURE 2-48: Eye-tracking heatmaps show how readers scan websites in an "F" pattern.

from there. This is why most websites place their logo in the top left rather than at the bottom.

On to hierarchy! When you're designing, some elements are more important than others. Your job is to establish a visual hierarchy so your readers and users can see which items are important. Essentially, visual hierarchy guides your audience and helps them decide what to look at first **(FIGURE 2-49)**.

Portions of this page are reproduced from work created and shared by Google and used according to terms described in the Creative Commons 3.0 Attribution License.

FIGURE 2-49: The left block has no hierarchy. On the right, the increased size of one element makes it stand out first.

A page without hierarchy appears flat and boring. There are several options for indicating importance and increasing visibility:

- **Size:** Larger items are more important and eye catching.

- **Color:** More saturated colors are more visible than low-contrast colors. Warmer colors (reds, oranges, yellows) are more visible than cooler colors (blues, purples, greens).

- **Positioning:** Items placed in the upper left corner have more importance.

- **Contrast:** Items with more contrast will be more eye catching than items with low contrast.

- **White space:** Items with more white space around them appear more important.

- **Typography:** Typeface selection can be another indicator of hierarchy. Using one font for your header and another for your content can create visual hierarchy.

Typography is particularly important for web work. The following paragraph has no visual hierarchy, giving every item the same importance:

> We're having a party! You're invited to the annual celebration of the Zuni Cafe. Celebrations will be held at Zuni Cafe, New York, at 6 pm. Formal dress required; please RSVP.

We can increase hierarchy by adding some white space:

> We're having a party!
>
> You're invited to the annual celebration of the Zuni Cafe. Celebrations will be held at Zuni Cafe, New York, at 6 pm.
>
> Formal dress required; please RSVP.

Once we start playing with typography and color, hierarchy becomes much more evident:

> ## We're having a party!
> You're invited to the annual celebration of the Zuni Cafe. Celebrations will be held at Zuni Cafe, New York, at 6 pm.
>
> FORMAL DRESS REQUIRED; PLEASE RSVP.

All three are readable, but the last example is the most visually interesting and makes the hierarchy of information immediately clear.

How can you tell whether you've properly added hierarchy to your design? Try the squint test. Squint at your design until the design blurs.

When the content disappears and you can only see the layout and elements abstractly, it's a lot easier to see what pops out first. On *New York Times* website (FIGURE 2-50), the logo and

FIGURE 2-50: A blurred version of the *New York Times* home page. When you can't read the text, content hierarchy becomes more clear.

images pop first (especially the orange button on the ad, which might not be ideal because it distracts from the website itself). In the content, the centered image and leftmost article with the largest headline text also stand out.

When you're working on a design, write a list down of all your elements and rank them in order of importance. Then look at your design and rank the elements based on how they look in the design. Do the two lists match up?

When we fail to establish visual hierarchy, the reader or user won't know where to start or end on your design; this confusion could reduce user engagement and conversions. The design won't work as well.

If something needs to look more important and higher in the visual hierarchy, play with the design and make some changes based on the ideas mentioned above—change the color or positioning, add spacing, improve the typography—until the item is more visually different.

If you'd like to break the "F" pattern in your design, you can do that by establishing proper hierarchy using the visual tools we listed above. For example, draw attention to an element at the top right of the page by making it larger or more colorful. The "F" pattern is a great starting point with natural hierarchy due to a typical placement of elements, and hierarchy is in our toolbox to break that pattern while still achieving a natural-feeling website.

Real-life examples

Let's analyze our widget to determine some of the ways that we're establishing visual hierarchy (FIGURE 2-51).

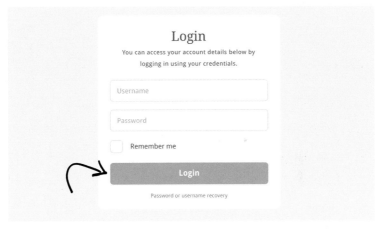

FIGURE 2-51: A brighter color for our download button raises its visual hierarchy. We don't want to accidentally hide this important button!

The first two important things we see are the headline and the login button. The headline is both a different color from the content and has a larger font size. The login button is a contrasting, warmer color than the rest of the widget. You'd probably see the headline or button first and then move through the rest of the widget—so we already established proper hierarchy when we updated the color and type in the earlier sections. Go us!

We went over a lot of these principles—color, typography, and white space—in earlier sections. Layout is where these concepts really start working together. Now we're going to move from visual design on its own to thinking more about user experience.

More about content and hierarchy in the following section, where we'll talk content principles!

CONTENT

AT FIRST GLANCE, A DISCUSSION ABOUT CONTENT might seem out of place in a book about design, but making good choices about words and content plays a huge part in ending clutter.

This section covers content principles and strategy to help you create content that users will read and enjoy and that improves how well your design works.

Dense paragraphs of content contribute to the feeling of clutter, especially when content is viewed on a screen (rather than in print). Studies have shown that online readers are more likely to skim content than read in full. Big chunks of text are more likely to be skipped than read **(FIGURE 2-52)**.

> People rarely read Web pages word by word; instead, they scan the page, picking out individual words and sentences.
> *How Users Read on the Web, Nielsen Norman Group*
> (*hellobks.com/hwd/45*)

First off, when it comes to content, less is generally more.

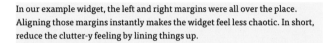

I take interest

In our example widget, the left and right margins were all over the place. Aligning those margins instantly makes the widget feel less chaotic. In short, reduce the clutter-y feeling by lining things up.

For web work, use shortcuts by using front-end web frameworks that include a grid, such as Foundation, Bootstrap, Skeleton, and PureCSS, which'll make it near-impossible to use random placement of HTML elements.

For non-web-design work and mockups, you can also use grids and guides within design programs like Sketch, Photoshop, and Gimp. Reduce clutter by limiting the colors in your design.

It can be super tough to choose colors, one of the reasons why color theory is often a semester-long class at design schools.

FIGURE 2-52: Users read web content fully only if their interest has already been piqued—otherwise, they skim content quickly as they look for something that grabs their attention.

It's tempting when writing web content to cover every possible detail and use big, professional words. Instead, it's preferable to write less and use smaller, clearer words that help folks understand what you're trying to say using the fewest number of words. Big paragraphs of text are an indicator of clutter— a good rule of thumb is no more than two to three sentences per paragraph.

When faced with a giant paragraph of information, try to shorten and simplify it. Let's look at a great example from *Simple and Usable Web, Mobile, and Interaction Design* by Giles Colborne (*hellobks.com/hwd/46*).

> Please note that although Chrome is supported for both Mac and Windows operating systems, it is recommended that all users of this site switch to the most up-to-date version of the Firefox web browser for the best possible results.

While the above is technically correct and covers all the bases, it's rather long and wordy. We can shorten:

> For best results, use the latest version of Firefox. Chrome for Mac and Windows is also supported.

Still correct, it covers everything we want to cover, but is much shorter and easier to understand at a glance.

It can be hard to limit yourself when it comes to content. I get it—you have so much important stuff to say! But remember: no matter how much you write, the reader will still only read 80, 100, maybe 200 words if you're *really* lucky.

Provide more content than the reader is willing to read, and they're likely to skim or abandon ship altogether.

If you write comprehensive paragraphs that bury the most important pieces of information deep inside, you're taking a big risk that the reader will waste their 80 words to read less important information.

The more content you present, the less control you have about what the reader actually consumes.

If you can't shorten something by using fewer words, you can also try breaking a paragraph into bullet points. Take this paragraph for example:

> We made a bunch of changes: The registration chapter has been split in two, between adding registration and then associating users with objects. The chapter was giant before, so this makes it more manageable. Screenshots of the admin have been updated to reflect the new Django 1.9 styles. The few minor typos have been fixed. Updated the version of django-registration-redux that we use to 1.3. Last but not least, the Introduction has been updated.

This paragraph would be better suited to bullet points:

> We made a bunch of changes:
>
> - The registration chapter has been split in two, between adding registration and then associating users with objects. The chapter was giant before, so this makes it more manageable.
> - Screenshots of the admin have been updated to reflect the new Django 1.9 styles.
> - The few minor typos have been fixed. Updated the version of django-registration-redux that we use to 1.3.
> - The Introduction has been updated.

Bullets and other visual aids help the reader pick out pieces of information from a block of content, making it more likely that the content will be read.

I also encourage strategic bolding. Especially in technical documents, it helps draw attention to the most important part of a sentence, further making your content easier to skim.

For example:

We made a bunch of changes:

- **The registration chapter has been split in two**, between adding registration and then associating users with objects. The chapter was giant before so this makes it more manageable.

- **Screenshots of the admin have been updated** to reflect the new Django 1.9 styles.

- **The few minor typos have been fixed.** Updated the version of django-registration-redux that we use to 1.3.

- **The Introduction has been updated.**

The most important parts have been bolded, so it's easy for readers to skim the content, find the parts that are most interesting to them, and read further if they like.

Note we also changed the formatting of the headline! Let's take a look at headlines in general.

One of the best ways to organize your content is by using headlines. These are parts of the text that are presented in a larger and more prominent style to introduce the content that follows, making the text easier to skim and read.

Headlines can suffer from wordiness—you also want these parts of your text to be as short and clear as possible to aid understanding and keep up excitement. For a better user experience, shorten and clarify your headlines, and also make sure that they're exciting and talk benefits for your reader.

For example, look at two versions of my website advertising my other books, the *Hello Web App* series. In FIGURE 2-53, the headline is wordy and boring. In FIGURE 2-54, the headline has

FIGURE 2-53: The *Hello Web App* website with a very wordy headline. Technically correct, but not very interesting.

FIGURE 2-54: This headline is still correct, but it's more exciting and shorter, Most important, it does a better job of drawing attention to the benefit of the reader.

been rewritten to be short and clear, talking about what the book *does* for the visitor of the website, making it more likely that the visitor will want to explore my website further.

If you're writing headlines that introduce a product or are designed to draw attention, make sure you write them in a way that showcases the benefits of the product rather than just describing what the product does.

Changes like this can drastically affect your bottom line. In a study by webprofits (*hellobks.com/hwd/47*) **(FIGURE 2-55),** changing the headline to talk about benefits rather than details increased the conversion rate (the number of people who bought the product) by 52.8 percent.

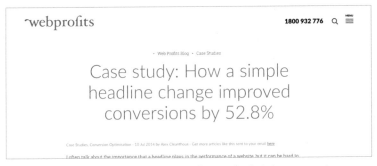

FIGURE 2-55: Changing a headline to talk benefits, not details, led to a conversion increase of 52.8% in this example by webprofits (*hellobks.com/hwd/47*).

Real-life examples

Back to our little widget! After learning about these content principles, we can update the content significantly **(FIGURE 2-56)**.

FIGURE 2-56: We've updated the content in our widget to make it shorter, clearer, more exciting, and easier to understand.

We've shortened the content and made everything more exciting and personable. The form itself has been updated to a more friendly, human-sounding sentence, and the button has been updated from a simple "Login" to a more actionable-sounding "Log into your account". All in all, these updates should make the widget easier to read, understand, and use at a glance.

Content might not initially feel like a part of design, but it really, really is huge. Content, writing, and words define how your design (especially web-focused design) will be consumed, so make sure that you spend some of your design time working on improving your content and writing.

In short, cut down and clarify your content as much as possible. Keep your paragraphs to two to three sentences.

Make sure your content work extends to your headlines as well. If your headlines introduce a page or product, make sure the headline talks about how the page or product affects the reader rather than what it does.

Next up, we're going to talk user experience!

USER EXPERIENCE

THIS IS THE SECTION THAT TIES EVERYTHING TOGETHER. All your work to make your designs enjoyable and easy to look at, your content short and easy to parse, your layouts easy to understand—all these concepts tie into your *user experience*.

User experience (UX) is the overall experience a user has when using your website—the most important part of design.

Let's break down the parts of user experience.

Before designing

Before you can even begin design work, you should be able to answer a few primary questions: Who is your ideal user? What is their personal goal? What are your business goals? User experience research can help answer these questions.

For example, someone visiting your personal website could be looking for more information about your work and what you do (their personal goal); your business goal could be to sign them up for your personal newsletter.

Or, for a web page for a new iPhone app, your user would be looking for information about your app and how to download it; your business goal would be to get as many downloads as possible.

We're looking at two things: whether the customer is achieving what they want to achieve and whether you're achieving your business goals. Hopefully your design achieves both.

You may not need to do user experience research for a small personal project, but there are some things you can do in the predesign phase of big, important projects to help figure out your users and their goals.

Competitive analysis

In a nutshell, look at your competitors and analyze what they do well and not so well.

Chapter 3, Section 3.1: Finding Inspiration goes over how to look at websites and pick out ideas and things you can implement in your own design. Looking at competitors is a crucial part of creating your site's user experience—it gives you a standard for comparison and helps you create a user experience superior to that of your competitors in order to lure their customers away.

Surveys and interviews

How will you learn what your customers want and need if you don't ask them? A lot of web designers fall into the trap of designing for folks they don't understand—for example, designing a wedding planning app if you haven't planned a wedding yourself. You might have some thoughts and ideas about what a person planning their wedding needs, but you may make some incorrect assumptions unless you actually talk to someone planning their wedding.

You want to have a good feel for what your ideal user wants and needs from your design. Having this knowledge will help you design better so the customer can achieve what they want, which should in turn allow you to achieve more of what you want.

During the design process

You don't want to spend the time implementing a design in code and then find out that folks are confused by a specific process or flow on your website (for example, the steps someone might take to buy an item). To avoid these missteps, you can do a "test run" of your design by using wireframes, prototypes, and usability testing.

Wireframes, prototypes, and usability testing

Wireframes and prototypes are fast, low-fidelity mock-ups of your design. Usability testing allows you to get feedback on layouts and interactions before you move forward with building the design in full. We'll examine these concepts in more detail in Chapter 3, Section 3.3: Prototypes.

Wireframes and prototypes allow you to build quick-and-dirty mock-ups of your design and take them to others to get their feedback, learning whether your layout and flow make sense. Research shows that you can find and address big usability issues with a low-fidelity paper prototype just as easily as you can with a high-fidelity screen prototype. Save yourself valuable time by testing your layouts and flows early on.

After launching your design

Once you've launched your design, you need to track whether you're actually meeting your goals. How is your conversion rate? Do you have a low enough bounce rate, the percentage of people who immediately leave a website after loading? Are you getting good customer feedback? To help answer these questions, you can conduct a variety of tests to measure this data.

More usability testing

A crucial part of user experience is making sure you use appropriate analytics packages (such as Google Analytics: *hellobks .com/hwd/48*) to collect data on how well your website is working after launch.

A/B testing is the process of testing a design change against your current website—such as testing two different versions of a headline and seeing which version better achieves your goal, such as increasing the number of purchases visitors make. These tests can be run after launch to continuously improve how well your designs are working.

Usability testing is the process of showing your designs to others to solicit feedback and ensure your website is usable by folks other than yourself, and we'll cover that in Chapter 3, Section 3.4: Getting Feedback.

Shortcuts

Now for my favorite part: the most important things to remember when thinking about user experience.

Make desired actions easy to find and use

Whatever you want your user to do, make sure the steps to complete that action are easy to find and straightforward to use.

It makes no sense for your submit button to be plain and hard to see—a bright button stands out and makes forms easier to submit (FIGURE 2-57).

submit me

submit me

FIGURE 2-57: Submit buttons for forms are important (as we talked about in the last section!) On the left, the button might be a bit too hard to find. The right button is easier to find and use.

Newsletter signup links shouldn't be hidden in the content (remember what we said about skimmers). Display them on their own line to make them easier to see (FIGURE 2-58).

I believe folks who know a bit about design will be more likely to hire a professional designer when the time is right. I'm a designer, and even I hire designers for things bigger than me. And if you agree, you should join my newsletter.

 vs

I believe folks who know a bit about design will be more likely to hire a professional designer when the time is right. I'm a designer, and even I hire designers for things bigger than me.

If you agree, you should join my newsletter.

email address   sign up

FIGURE 2-58: If you want someone to sign up for your newsletter, don't hide the link in your content. Push it out and make it visible.

Pay attention to website size

Slow downloading will prompt people to leave a website without seeing it. Pay attention to image size; with large screens and fast downloads the norm at home, it's easy to forget that a lot of folks still browse websites on cell phones with poor reception or coffee shops with overloaded Wi-Fi. Also consider JavaScript; downloading lots of scripts will slow a website down, as will slow-performing scripts.

Check your website in multiple browsers using multiple sizes and multiple download speeds to make sure it loads quickly.

Run usability tests

We'll cover usability testing in more depth in Chapter 3, Section 3.4: Getting Feedback. Essentially, show your designs to others and get their feedback, as hard as that might feel. As designers, we're naturally blind to problems in our design, so it's crucial to get more eyes on it to find those issues.

Include analytics

Make sure to include analytics-tracking software in your designs so you can see how your design is performing after launch. At the very least, consider Google Analytics (*hellobks .com/hwd/48*) or a customer data platform such as Segment (*hellobks.com/hwd/49*) that you can tie in with Google Analytics and other analytics platforms.

Real-life examples

Going back to our widget (**FIGURE 2-59**), we can see that our login button is a bright, easy-to-see color. Users aren't going to be confused about how to submit!

User experience is one of those things that's hard to show in a book. Remember that you should do your best to try to understand your audience, pay attention to your pages and how a person moves between them (and make sure that's easy to do), get feedback from your designs from outside sources, and pay attention to how well your design is doing after launch.

Congrats on finishing this section!

FIGURE 2-59: User experience goes hand in hand with hierarchy. Our form is more usable because our submit button is in a bright, contrasting color.

IMAGES AND IMAGERY

SO FAR WE'VE BEEN WORKING SOLELY WITH LINES, boxes, and text for your designs, but images (and imagery such as icons and graphics) can play a huge part. In this section we'll address where you can find images to use for your project and how to use them effectively.

First off, let's cover royalties and licensing. As easy as it is to right-click and "save as" almost any image you find on the internet, you can't use just anything you find. Images have copyright protections, and reusing an image created by someone else without permission is illegal.

An overview of terms you should keep an eye out for:

- **Rights-managed images:** Should include additional terms indicating allowable uses for the image.
- **Royalty-free images:** Generally you can use the image *as long as* you correspond to the terms of the image. For example, a lot of royalty-free images can be used only in not-for-profit contexts, meaning you cannot use them for any application in which you could potentially make money.

- **Creative commons:** A system created for content creators to allow their work to be used without requiring express permission. Creative commons images can be licensed in many different ways, from requiring attribution, denying derivatives (you can't alter the image yourself), prohibiting commercial use, or allowing any use.

If an image doesn't specify that you can use it, then *you are not allowed to use it for your project*. The vast majority of the world has adopted the Berne Convention, which states that copyrights are granted to the author by default—even if not specified. You don't actually need to put little copyright notice footers on your website!

Fundamentals

Let's cover some quick fundamentals to remember when you're adding imagery to your designs.

Don't feel like you're required to use photos or icons

Don't fret that you *need* to use photos of people in your designs. Type and screenshots of your product can go a long way (FIGURE 2-60).

By playing with typography—using different fonts and styles—you can create a beautiful and clean website without needing to fuss with images.

If you're building a product for the web, the website for that product doesn't need much more than screenshots (FIGURE 2-61).

You can always add images and imagery later—starting out with a clean, text-based website could help you launch faster without nitpicking over these details.

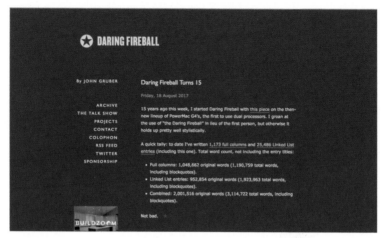

FIGURE 2-60: John Gruber's Daring Fireball website is a beautiful, successful design using all text (other than the icon in the logo).

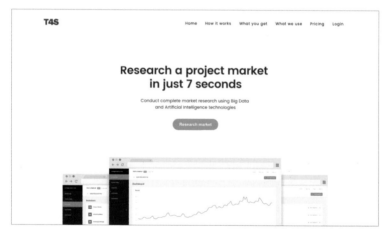

FIGURE 2-61: The T4S website is clean, featuring screenshots of their product.

Faces are powerful

Happy faces provoke happy emotions; angry faces provoke angry emotions. You can use a face in your design to emphasize the emotion you want and draw the eye to the most important part of the page **(FIGURE 2-62)**. This is a powerful tool in your design toolbox.

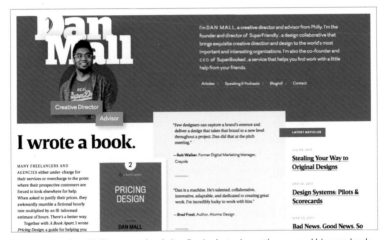

FIGURE 2-62: Dan Mall's personal website. Dan's photo draws the eye, and his gaze leads us to the introduction text.

Pay attention to file size

Images can easily bloat a website's size very quickly, leading to slower download times and frustrated users who don't stick around. Additionally, more screens are going retina with higher resolutions, with smaller pixels leading to crisper more crisp displays—making photos and images at the old resolution look blurry.

Some principles to keep in mind:

- **Make sure your images are the maximum necessary size.** No sense in putting a 2000px image into a 1200px box.

- **Provide retina and non-retina images.** Set up your HTML and CSS so folks using a retina screen can download the more detailed, higher-resolution images, and those on a traditional screen will get the smaller images with no wasted pixels. This is a great guide on doing so: *hellobks.com/hwd/54*.

Beware of stock photos that look like stock photos

Searching for "stock photos" on Google will get you to multiple websites that purport to offer a lot of great options you can use (paid and free), but there are a whole lot of low-quality images. Take a look at some of the giant websites like iStock **(FIGURE 2-63)**, but try to find images that feel more natural than the example image (digital construction in the future!).

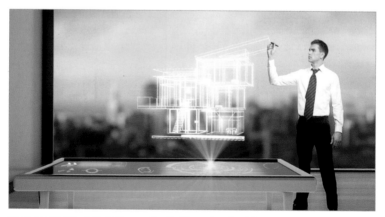

FIGURE 2-63: iStock (*hellobks.com/hwd/53*) has a lot of stock images to go through, but beware of images that feel overly posed and unnatural.

Icons

While images are great for backgrounds and big elements, icons are more like supporting characters—little graphics that support your content and design. You don't *need* icons, but they can make your design feel unique.

Icons support information absorption by creating a graphical representation of content. The Kite website uses icons effectively to draw the eye to the three columns of features using abstract representations of the content **(FIGURE 2-64)**.

FIGURE 2-64: Kite's website icons are simple but effective.

Of course, this is also a place you need to pay attention to licensing. Just like photos, icons are copyrighted.

Shortcuts

Stock photo websites

My favorite resource for natural-looking stock photos is Unsplash (*hellobks.com/hwd/56*) **(FIGURE 2-65)**, which features free Creative Commons images.

Other stock photo websites include IM Free (*hellobks.com /hwd/60*), picjumbo (*hellobks.com/hwd/61*), iStock (*hellobks.com /hwd/62*), Gratisography (*hellobks.com/hwd/63*), and PhotoPin (*hellobks.com/hwd/57*) **(FIGURE 2-66)**.

Imagery (icons, graphics, and illustrations)

To add that little special touch to your designs using icons, try either websites with predesigned icons, such as The Noun Project (*hellobks.com/hwd/58*) **(FIGURE 2-67),** or get them custom designed for you through websites such as Fiverr (*hellobks.com /hwd/59*) **(FIGURE 2-68)** and Upwork (*hellobks.com/hwd/64*).

FIGURE 2-65: Unsplash (*hellobks.com /hwd/56*) is a great source for natural-looking images.

FIGURE 2-66: PhotoPin (*hellobks.com /hwd/57*) searches Flickr for Creative Commons images.

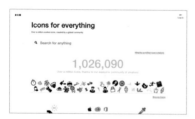

FIGURE 2-67: The Noun Project (*hellobks .com/hwd/58*) includes lots of different Creative Commons icons you can use in your projects.

FIGURE 2-68: Need an icon designed? Try something like Fiverr (*hellobks.com /hwd/59*)

Real-life examples

Our widget was already readable, well spaced, and usable—but by adding some polish through a background pattern, a subtle drop-shadow, and a fun icon, we can really elevate its appearance (**FIGURE 2-69**). We can often get away without these little details (and I often personally skip this effort on some of my own quick design work), but as you can see, it can be time well spent on important design work.

FIGURE 2-69: Our widget was fairly good already, but adding a background pattern, a light drop-shadow, and a fun icon takes it to another level.

Next up, we're going to cover some random and fun design tidbits that didn't fit into these earlier sections.

EXTRA TIDBITS

LET'S WRAP UP THIS CHAPTER by going over some fun design tidbits. This section is a catchall for some things we covered a bit of in earlier sections and some that didn't fit perfectly in one category; all should be interesting and will add a bit of spice into your design work.

Start simple

When you're starting out, deliberately choosing simple interfaces and designs will make your life easier. Simple layouts, simple user interface (UI), and simple UX might make you feel boring (you're not!), but they are effective and easier for newbies to wrangle (FIGURE 2-70).

Don't be ashamed to start out small—after all, you can always add more elements or redesign after your first iteration.

Consider the "rule of thirds"

The "rule of thirds" is a concept usually applied in photography. Centering an object in a photograph makes the

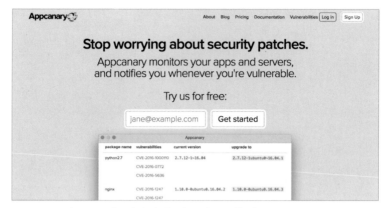

FIGURE 2-70: Appcanary's website is simple and very effective.

photograph feel boring. Instead, divide the image horizontally and vertically into thirds; the intersections of these lines are great guidelines for placing points of interest, creating photos that are visually more interesting (**FIGURE 2-71**).

This also can apply to web design—one of the reasons why a three-column website design is visually appealing. You can also place your background images and important features (such as faces) along these lines (**FIGURE 2-72**).

Again, this is another rule that can (and often should be) broken, but it's a good concept to keep in mind.

Avoid pure black

Pure black—which doesn't occur in nature—can make your design feel stark and unnatural. Instead, use color that is not quite black; in hexadecimal, something like #222222 is great for headlines and #444444 for text, rather than #000000. You can also use the specific CSS property RGBA (color:

FIGURE 2-71: The rule of thirds. Points of interest aren't centered—rather, points of interest are placed along the "thirds" intersections for more compelling layouts.

FIGURE 2-72: Comovee's website background basically aligns with the page in thirds.

`rgba(0,0,0,0.2);)` so your text darkens the background color rather than using pure gray **(FIGURES 2-73 AND 2-74)**.

White is less likely to invoke this feeling of starkness, but it still could be worthwhile to play with using almost-white in your designs.

Lighting should come from above

If you're adding shadows or gradients to your design elements, set it up so that it looks like it's been lit head-on or from above **(FIGURE 2-75)**. Objects lit from below look unnatural, as we're used to light normally coming from above (the sun, overhead lights, and so on) **(FIGURE 2-76)**.

A good rule of thumb: Set your minimum to half (or less) of what you actually would like to raise.

My Kickstarter's minimum was $15,000, which tells you that I was really aiming for $30,000. So while my campaign looks wildly successful, it actually hit less than I was aiming for.

A couple reasons why you should aim your campaign for less than you want:

- With Kickstarter, **you don't get your cash unless you hit your minimum**, so it would suck to raise less than you hoped *and* not get anything.
- **A lot of folks will only jump on already-successful campaigns**, so by

FIGURE 2-73: Black is readable but stark and a bit hard on the eyes.

A good rule of thumb: Set your minimum to half (or less) of what you actually would like to raise.

My Kickstarter's minimum was $15,000, which tells you that I was really aiming for $30,000. So while my campaign looks wildly successful, it actually hit less than I was aiming for.

A couple reasons why you should aim your campaign for less than you want:

- With Kickstarter, **you don't get your cash unless you hit your minimum**, so it would suck to raise less than you hoped *and* not get anything.
- A lot of folks will only jump on already-successful campaigns, so by

FIGURE 2-74: Dark gray is still readable but a lot easier to look at on a computer screen.

FIGURE 2-75: Lit from above, the button feels like it's popping out naturally from the page.

FIGURE 2-76: The button looks unnatural when lit from below.

Contrast highlights what's important

In your designs, widgets, and forms, you can play with contrast to make something stand out (or stand back). This works particularly well with form elements, when you want to visually indicate what is editable versus what is just help text **(FIGURE 2-77)**.

Colors look different depending on the environment

The lovely peach in your color palette might suddenly look brown when placed on a bright orange background **(FIGURE 2-78)**. It's not your palette, it's our eyes—our perception of color changes depending on the environment. Feel free to alter the colors on your palette a bit so they visually match their source, even in a new environment. Your color palette doesn't have to be set in stone!

FIGURE 2-77: The form's help is lighter compared to the main text, indicating that it's less important.

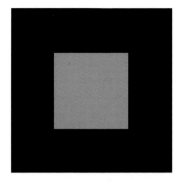

FIGURE 2-78: The inner square is the same color in both images, but looks significantly darker and browner in the example on the right.

Be careful about overlaying text on images

We went over typography, readability, and images in separate sections. Combining images and text can be hard to do because the image itself can hinder readability (and ergo, how well your design works for the user) **(FIGURE 2-79)**. Try overlaying transparent color on top of an image (black is typically used, but you could also try white or colors) **(FIGURE 2-80)**, adding a solid block of color behind the text, or blurring the image behind the text if readability seems like it'll be a problem.

FIGURE 2-79: Text can become unreadable depending on the background.

FIGURE 2-80: Darkening the background ensures that text will always be readable.

Some tips when designing for repeated workflows

This book mostly focuses on designing web pages that try to convert a first-time visitor into a certain action, which means we're trying to avoid designs that require any kind of learning curve.

However, internal tools that will be used many times every day allow us to make a tradeoff between requiring some upfront learning and making our users' workflow more efficient. Examples include creating keyboard shortcuts or allowing more clutter in order to make vital information available in one place. The Bloomberg Terminal—borderline terrifying for novices but irreplaceable for professionals—is one example of purposeful clutter in expert interfaces (FIGURE 2-81).

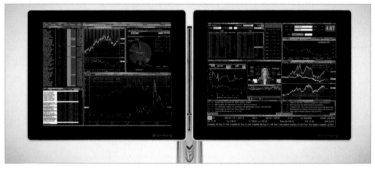

FIGURE 2-81: The Bloomberg Terminal design is cluttered and chaotic, but with purpose.

In an ideal scenario, we could create the best of both worlds: a design both intuitive and beautiful that lets users perform any task as fast as possible. But reality dictates that we usually need to compromise on either simplicity or function. The decision about which to favor must be made consciously.

And that's the end of the chapter on theory and shortcuts!

Over the last few sections, our widget has slowly improved.
Here's how we've transformed it (FIGURE 2-82).

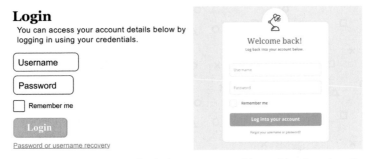

FIGURE 2-82: Where we started and where we are now with our widget. In each section we added improvements, creating a more beautiful and more usable widget.

Pretty awesome!

CHAPTER 3
THE PROCESS AND TRAINING YOUR DESIGN EYE

NOW THAT WE'VE CHATTED ABOUT DESIGN PRINCIPLES and the pieces of design, let's work on the process of design. While the last few sections have thrown a lot of theory at you, these next sections are geared toward adding real-world context to what you have learned and helping you do design work.

The following sections will walk you through the parts of design from start to finish—finding inspiration, sketching your project, getting feedback, and building. We will talk about these steps in abstract but also apply them to a sample project so you can see how everything comes together.

For our sample project, we're going to build on a theoretical home page for an open source project—a place where visitors can learn about and use the project, with links to documentation and information on how to contribute to the project.

Let's get started!

FINDING INSPIRATION

WHAT'S THE FIRST THING YOU SHOULD DO when starting a design project?

If you think that you should sit down and start working on ideas, sketching layouts, planning your content—you're close, but not quite ready to go there yet.

Sitting down and immediately working on a design problem without first doing research and looking for inspiration is like trying to code without access to the internet to look up questions and errors—doable, but slower and more frustrating.

Think of inspiration as visual debugging. Looking at great design and inspiration will help you figure out solutions to problems that you may run into when creating your own designs.

Where to find inspiration? For web design, there are tons of websites that collect and share good designs (FIGURE 3-1).

Of course, do *not* rip off designs. If you find something you like, you can do something similar, but ripping off something verbatim is a very big no-no. But you *can* be inspired by layout,

The Best Designs (hellobks.com/hwd/67) *Unmatched Style (hellobks.com/hwd/68)*

Awwwards (hellobks.com/hwd/69) *Site Inspire (hellobks.com/hwd/70)*

FIGURE 3-1: Some of the many options you can use to find web design inspiration.

the treatment of color, the tone, the imagery, and the typography and implement the things you like into your own project. Try to focus on a specific aspect of a design that you like, rather than the entire thing, and apply it to your project.

Your job as designer isn't to reinvent the wheel. Noting established conventions and things that "work" will help you implement familiar flows and layouts in your own designs—making it more intuitive for your users to navigate your product.

My *Hello Web App* books were inspired by the wonderful *A Book Apart* books (FIGURE 3-2).

I'm particularly proud of my design for *Hello Web App*, but I sincerely don't think they would have turned as they did without being heavily inspired by *A Book Apart*. Seeing those books helped me determine what I liked in terms of size and

FIGURE 3-2: My *Hello Web App* books on the left, and the beautifully designed *A Book Apart* series on the right.

thickness, and the flat, bright colors inspired my own graphic treatment of the *Hello Web App* cover.

Pablo Picasso famously said, "Good artists copy, great artists steal." When you find something you love, something that you'd like to do yourself—whether it's layout, color, font choice, and so on—use that as inspiration to build something *similar* but not *exactly* the same.

One of the most useful things you can do to make yourself better at design is to look at other designs and think critically about what they do well and what they do poorly. This is a great process to practice while you're looking for inspiration!

The more you do this (with good designs as well as bad), the more you'll train your "design eye" and design intuition, making it easier for you to create good designs from scratch.

Let's use the GitHub home page design as an example **(FIGURE 3-3)**.

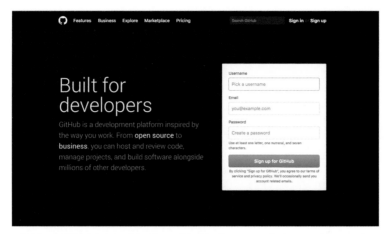

FIGURE 3-3: GitHub has consistently great-looking home page designs, great for practicing picking out good design decisions.

Using the principles we went over in Chapter 2, Theory and Design Principles, what would you say they're doing well? Some of the details I've picked out:

- **Lots of white space.** The headline, content, and form have plenty of white space around the top and bottom, drawing the eye and emphasizing the content. Just enough white space—not too much and not too little.

- **Sign-up form is front and center.** No need to jump through hoops—GitHub makes it super simple to sign up for an account right from the front page. Note that the "Username" field has focus by default so the user can start typing immediately.

- **Important words are bolded and linked in the content.** Both "open source" and "business" are linked, bolded, and in a brighter color. At a glance you can see the two areas that GitHub focuses on.

- **Subtle background pattern helps the white space feel less empty.** Imagine the background in a flat gray—the design would still look good but would feel less polished. The subtle pattern fills the white space while letting the content shine.
- **Main button is a bright, stand-out color.** The majority of the design is dark and gray, making the bright green form submit button stand out.

What other good design decisions can you see?

Start browsing the web more mindfully, and think critically about what designs are doing well and poorly. Over time, your design intuition will improve. Making this a practice will make it easier for you to create better designs naturally.

This is especially important when you're actively working on a new design. Start out by looking for inspiration in examples of good design in your field and by looking at your competitors. Pick out what they're doing well, and implement those ideas in your own design. Pick out what you think isn't working well (especially in your competitor's designs), and make sure you avoid those things in your own design.

Our sample project for Chapter 2 was our widget. For this chapter we're going to be building a home page for an open source project. Let's start by finding some really clean and visually interesting home pages for other open source projects (FIGURE 3-4).

There are a number of trends that will be nice to implement in our own design:

- **Lots of white space**—open source project home pages are fairly simple (compared to, say, a newspaper home page) and an airy, designer-y feel is a nice contrast to something aimed at developers.

Gulp

Rouge

Travis CI

Mocha JS

FIGURE 3-4: A few well-designed open source project home pages that we'll use for inspiration.

- **Bright colors**—either as a full background or as highlights on top of a white background.

- **Clear, prominent headlines that explain the tool.**

- **Code is front and center**—naturally, a product made for developers should show the code involved right at the start rather than hiding it.

This part of the process is very much affected by personal preference—I chose four designs that I personally loved and inspired me; you might disagree with my choices. That's totally okay! Design is very personal, and at the end of the day you will want something that represents *you*.

PLANNING

YOU COULD JUMP STRAIGHT INTO CODING and start creating your site using your inspiration, but the middle process of design—creating a plan, sketching ideas and layouts, and making a mock-up—will help you try more ideas and create a better design in overall less time. It's a bit of effort up front that'll save a lot of time down the line.

Your life will be so much easier if you do some basic planning before moving ahead with trying to sketch and develop your design. How many pages do you need? What kind of content do those pages need? What kind of forms do you need, and how many fields?

We're going to build our open source project home page sample project as a single page with links out to external documentation—making our plan rather simple. Just one page!

On this one page, let's list the elements we'll need:

- Logo/name of project.
- Headline explaining the project.

- Menu with links to external documentation on Read the Docs, GitHub page, author's Twitter.

- Block showing code (to show how easy it is to install).

- Three blocks with benefits and features in more depth.

- Positive quote from contributor.

- Repeated menu in the footer.

Most websites won't be this simple. Let's think about another scenario, such as a personal portfolio website for a designer named Jane, which would require more pages.

First, let's determine what pages we need for a personal website for Jane:

- **Home page:** The first page someone sees when they hit the website.

- **About page:** A deeper dive into Jane's background and experience.

- **Portfolio page:** An overview of Jane's projects.

  - **Individual project pages:** These pages will probably use the same layout, so we can group them under one category.

- **Contact page:** A page with Jane's location and information about how to contact her.

Right away, we can see that we have five basic types of pages that we will need to design layouts for. Now that we have the basic pages, what do we want each page to have on it?

- **Every page:**

  - Personal logo or name.

- Top menu listing all top-level pages in the website (that is, we're only going to list the four main bullets above, not each individual portfolio page).

- Footer menu after content.

- **Home page:**

  - Short, exciting description of Jane.

  - Photo of Jane.

- **About page:**

  - Longer descriptive paragraph.

  - Different image of Jane.

- **Portfolio page:**

  - Blocks for each of Jane's projects with a photo/screen-shot of each, project title, and links to individual project pages.

- **Individual project pages:**

  - Image representing project.

  - Long description of project.

  - Links to external resources (code on GitHub, launched website, and so on).

- **Contact page:**

  - Short introduction paragraph.

  - Email address and location.

  - Form for contact with fields for name, email, short content section, and a submit button.

Making a plan and listing of all these pages, features, widgets, and blocks will help you organize your design process and ensure that you don't miss an important element when sketching and building your design.

At this point, you could jump over to coding your site, but I highly recommend taking some time to build some prototypes. We'll cover this in the next section!

PROTOTYPES

THE PROCESS OF SKETCHING YOUR IDEAS and making proto-types will help you play with solutions and try out different ideas faster than if you moved straight to coding. This usually happens in tandem with the inspiration process. As you go through and find ideas you like, you should make note of these ideas you'd like to implement and sketch out how they could fit into your own design.

Sketching can seem more intimidating than it is. Perhaps you've seen sketches of websites that look like FIGURE 3-5.

You *can* sketch something like the one in FIGURE 3-5 if you like, but that really isn't how initial sketches are made. Sketching is more abstract, uses minimal detail, and works with your imag-ination to make everything come together. All you really need are boxes and lines to start jotting down some ideas for layouts and how your design should look (FIGURE 3-6).

This creates a quick and easy shorthand to start sketching out your design ideas. Here are some more examples from when I was redesigning the *Hello Web Books* website from my own sketchbook (FIGURE 3-7).

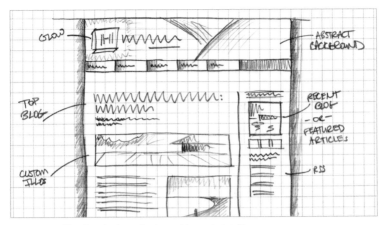

FIGURE 3-5: Sketches don't *need* to have this much detail.

FIGURE 3-6: Low-information sketches. Squiggles represent headlines or larger text, lines represent content, boxes indicate buttons, and x-ed out boxes represent images.

FIGURE 3-7: Messy (digitized) sketches from my personal notebook where I was planning the new *Hello Web Books* website design.

Again, these sketches aren't pixel perfect: they're meant to provide quick layout ideas to help you start building your design. Quick, messy, easy, and definitely not perfect.

You can upgrade your sketches by adding shading or different shades of gray, but try to keep your sketches as unstyled as possible in this initial round. No fancy fonts, no colors, no exact dimensions. Keep this round as easy as possible on yourself so you can quickly create as many different ideas as possible (FIGURE 3-8).

FIGURE 3-8: Paper sketches showing basic lines, boxes, image locations, and content placement. Sketches don't need to be too complicated.

Let's return to our sample project and sketch out a few quick ideas for home page layouts for our open source project (FIGURE 3-9).

Here are quick sketches of two different home page layout ideas, but I recommend that you do more when you're sketching

FIGURE 3-9: A couple of quick sketches showing two different home page layout ideas for our sample open source project home page.

your ideas. The more layouts you do, the more ideas you'll go through, which will help you gain confidence in the sketch (or sketches) you choose to build wireframes from.

If don't have a particular project in mind, sketching the layouts of your favorite websites is great practice.

Wireframing

Next, we make our quick sketches more concrete. Wireframing helps us build in extra detail from our sketches, working with real sizes for placement and helping us figure out spacing in a more refined representation of our design.

There are a lot of different software solutions available to help you with this stage (FIGURE 3-10).

GIMP *(hellobks.com/hwd/77)*   Sketch *(hellobks.com/hwd/78)*

Balsamiq *(hellobks.com/hwd/79)*   UXPin *(hellobks.com/hwd/80)*

FIGURE 3-10: A few of the software solutions available for wireframing.

Free Options

- **GIMP.** Open source image editor: *hellobks.com/hwd/77*
- **Inkscape.** Open source vector graphics editor: *hellobks.com/hwd/81*

Paid Options

- **Adobe Illustrator or other Adobe products.**
  hellobks.com/hwd/82

- **Sketch.** *hellobks.com/hwd/78*

- **Presentation software such as Keynote (Mac) and PowerPoint (Windows).** You can set up clickable areas that'll go to different slides or "pages" of your site, so your mock-ups can feel more interactive: *hellobks.com/hwd/83*, *hellobks.com/hwd/84*

- **Balsamiq.** Online wireframe software: *hellobks.com/hwd/79*

- **UXPin.** Online design platform: *hellobks.com/hwd/80*

Wireframes come between the sketch and mock-up stages. At this stage we are *not* choosing colors or finding fonts—we're just continuing to think about layout and flow.

Going back to our open source project home page, we're going to take my favorite sketch and use it to create a wireframe (FIGURE 3-11).

We're using a three-column layout, starting to plan our content (making sure it's short, easy to understand, and exciting) and figuring out spacing and layout. Once we have these elements in place, we can start testing our design (see Section 3.4: Getting Feedback) and trying out ideas until we have something we want to move forward with.

We're still missing major design decisions such as fonts, color, and graphics, but this bare-bones wireframe gives us a great place to start when getting feedback from others and coding the site.

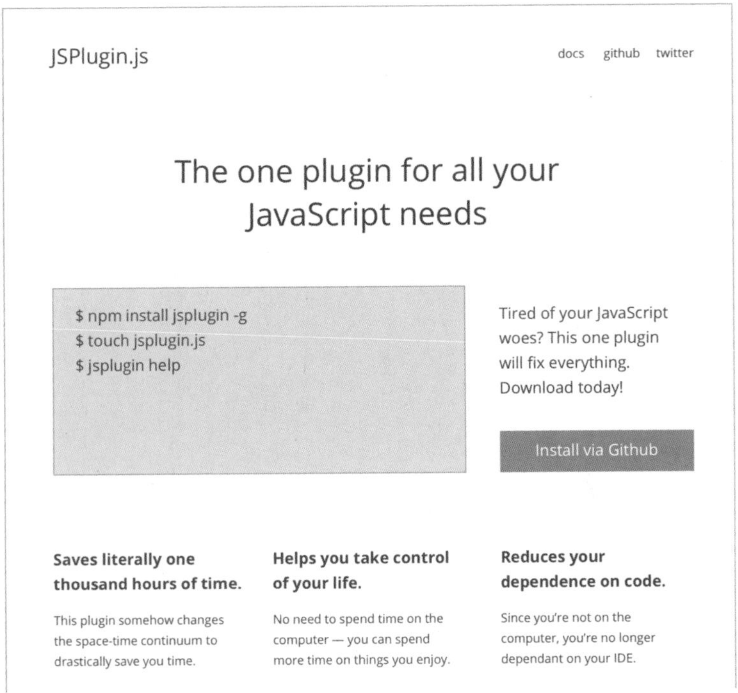

JSPlugin.js

docs   github   twitter

The one plugin for all your JavaScript needs

```
$ npm install jsplugin -g
$ touch jsplugin.js
$ jsplugin help
```

Tired of your JavaScript woes? This one plugin will fix everything. Download today!

Install via Github

Saves literally one thousand hours of time.

This plugin somehow changes the space-time continuum to drastically save you time.

Helps you take control of your life.

No need to spend time on the computer — you can spend more time on things you enjoy.

Reduces your dependence on code.

Since you're not on the computer, you're no longer dependant on your IDE.

FIGURE 3-11: A quick wireframe based on one of our sketches. No major design choices—just layout and spacing.

It's important to use real content—avoid fake content such as lorem ipsum (dummy text used by the design industry)—as much as possible. With real content, you'll know the length of text you need, which is crucial when making your layout, rather than using fake text that may end up being longer or shorter than the text you eventually use.

Pay attention to the content principles we've gone over, making sure your content is short, easy to read, and exciting. As you go through the sketching, wireframing, and mocking up stages—revising and testing different versions of your designs—you should be revising and improving your content as well.

Let's look at an unrelated idea so we can get another example of what wireframes would look like (FIGURE 3-12).

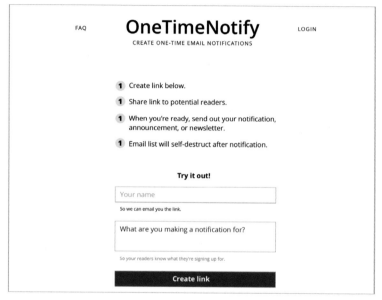

FIGURE 3-12: Another wireframe example.

Again, no fonts or colors, but this quick wireframe shows planned layout (mainly one narrow column with menu options around the logo), and some design decisions, such as using a bulleted list for content and a form that's high enough on the page to be seen without scrolling.

Wireframing should help you make simple layout decisions so you can try ideas quickly and test them without spending time on small design details or changing large chunks of code.

Depending on what you're doing, you *could* move straight to code after creating your wireframes, or you could create a mock-up to explore final colors, fonts, and images.

Mocking up your design

Wireframes are pretty low-fidelity—we can also create high-fidelity mock-ups in our design programs before we move to code so we can get every design detail planned (fonts, colors, backgrounds, and so on) before moving onto code.

This part of the design process is a bit tougher if you haven't used a design program before. In Chapter 5, Additional Resources, I cover tutorials and videos on basic design skills for major design programs.

Some pros and cons to spending the extra time mocking up your design:

Pros

- You'll see *exactly* what your design will (hopefully) look like when you move into building it in HTML and CSS.

- Objects can be easily moved and transformed, so you can change your mind more quickly than if the design were already coded.

Cons

- Adds an extra step that could take more time.

- As you're designing for the web, your eventual website needs to be *responsive* (the design adapts so it looks as good in smaller screens as it does in larger). Responsive designs are harder to mock up, as most design solutions are static. You'd need to create mock-ups that show the design adapting to different sizes (mobile, computer, and so on).

- If you're not comfortable with design programs (or you're more comfortable with code), mocking up may take more time than coding in HTML and CSS.

Your mock-ups should include all of the decisions you made before about content and layout from your wireframes with the additional bonus of planning for graphics, colors, fonts, and extra-beautiful design tweaks (FIGURE 3-13).

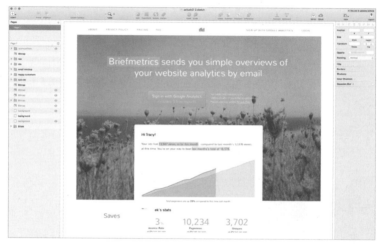

FIGURE 3-13: Mock-up showing color, font, and imagery decisions using the design program Sketch.

Let's take the wireframes for our sample home page and start adding some color, fonts, and imagery. Focus on making this first draft *just* good enough to launch and show others, as you can continue iterating and improving the design over time (FIGURE 3-14).

FIGURE 3-14: Back to our sample project. Go from wireframe to mock-up by adding color, imagery, fonts, and other style decisions. This example is still very simple, but it works as a first version. You could launch this and then iterate on the design and make improvements over time.

Once you have a mock-up you like, it's time for another round of testing and sharing the design with friends and family (covered in depth in the next section) to ensure you're not missing anything crucial before spending time coding the site.

At this point, you can copy your home page design and build mock-ups for the other pages (such as an About page), as well as mock up what your website will look like on smaller screens **(FIGURE 3-15)**.

FIGURE 3-15: The mock-up we created, altered to show how the design would look like on a smaller screen. Remember that it's important that your websites be usable and readable on smaller screens.

In the next section we're going to cover a very important part of the design process (hinted at throughout this section)—getting feedback!

GETTING FEEDBACK

ONE OF THE HARDEST BUT MOST CRUCIAL PARTS of the design process is getting feedback on what you've designed.

Unlike programming, there no simple "working" or "not working" answer to what you've designing. You can't tell at a glance whether your design is successful without getting feedback from outside sources. This simple step can be the toughest, because we're all naturally afraid of negative feedback. What if the person hates what we've designed? What if they find problems?

A design might feel right to you, but you're the designer—you know all the ins and outs of the design, how it works, and how it should be used *because* you're the designer. You'll only know whether those paths and objectives are obvious after showing your design to someone else. Your design might be beautiful, but you don't know whether it'll be appreciated and used until you show it to others.

It's natural to feel bad if someone discovers something that isn't working. When this happens, look at it as a learning experience and try to remember that these pieces of feedback will improve your design and teach you something new. Accept bad feedback with grace—there is no need to feel embarrassed or insulted. Better to hear bad news early on rather than after you launch.

It's totally normal to not want negative feedback, but finding out where your design can be improved will make your design stronger and your experience better in the long run. Detach your feelings, steel your will, and find the problems early.

Seeing your design with fresh eyes

Before you take your design to others, give yourself a chance to critique your own design. Simply taking a break from your design (a few hours or overnight) can help you figure out potential design issues and problems. This is the easiest way of getting feedback, because you're working with someone naturally agreeable—yourself!

When you come back to your design, try to distance yourself from your designer-self and to embody the potential user. Is your design usable? Easy to read and understandable? Is the design easy to use and natural feeling? Go through the user experience and imagine how the website will be used—find all the "easy" issues as early as possible before you share your design with others.

Another trick for getting a quick fresh look at your design is to take a screenshot and flip it horizontally. It's still the same layout, but your brain will lose some of the familiarity with the design so it will be easier to be more objective and pick out problems.

Some advice on showing your design to others

You could take your design to someone else, get a "looks great!" in response, pat yourself on the back, and move on with your day. But was that really valuable feedback?

Make sure that you're clear about what kind of feedback you're looking for. Maybe you'd like help finding problems with your design and you're open to general feedback (positive or negative) and criticisms, or perhaps you need help choosing between two options. Alternatively, you might be 90 percent done and only looking for minor tweaks before you publish.

Give your reviewers careful instructions about what you need and plenty of time to look at your work. A quick glance at your design isn't enough, so don't surprise the reviewer between other tasks. Let them have time to really think critically about what you have done. Quick responses are more likely to only cover superficial issues or be hollow praise.

If you receive only positive responses, try asking specifically for what the reviewer *didn't* like. This could prompt a more thorough response and also indicates that you're looking for (rather than avoiding) negative feedback.

Keep in mind that when you ask for feedback, you also need to judge the feedback that comes in. Talk to more than one person, as a problem that comes up only once might not be an actual problem. If an issue surfaces over and over, *then* you've found something that needs to be fixed.

Not all feedback has to be implemented. It's up to you, the designer, to figure out the true issues versus the things the reviewer mentioned because they wanted to be helpful.

Showing your design to friends and family

Your friends and family are the friendliest folks you can go to for design feedback, for better or for worse. Those that are closest to you are the most likely to want to please you and may want to give only positive feedback. Here is where it's especially important to make it clear you're looking for all feedback (positive *and* negative).

It's great if you can show your designs to folks who match your ideal user (if you're building a tool for developers, another developer friend would be a great tester), but don't feel limited to that rule—UX issues could be found by anyone.

Showing your design to strangers

Although significantly harder, showing your design to strangers can yield more valuable feedback because they aren't affected by their personal relationships with you.

Where can you find strangers who might be willing to give feedback?

- **Hackathons, sprints, and other community events.** Events and gatherings of like-minded peers are great places to get feedback on design work. Ask someone politely if they'd be willing to spend 5 to 10 minutes reviewing your design.

- **Online communities.** Forums and communities such as Reddit's *design_critiques* subreddit (*hellobks.com/hwd/85*) and Bootstrapped.fm (*hellobks.com/hwd/86*) can help review mock-ups and early versions of your design. Internet anonymity creates a tendency for strangers to be negative, so here especially you need to sort the useful feedback from the bad. A thick skin is essential.

- **Online review services.** There are quite a few online services that allow you to upload a screenshot of your design to get feedback. One example is Five Second Test (*hellobks.com/hwd/87*).

- **Folks at coffee shops and the like.** No hackathons near you? You can also try the classic usability test trick—buy a couple of $5 gift cards to a coffee shop of your choice, approach friendly looking strangers, and ask them for five minutes of feedback in exchange for the gift card (or a coffee, or something similar). Approaching a stranger in a public place can be scary, but getting a stranger's feedback on your design is so valuable that it's worth it.

This step is so easy to skip, especially when you're adhering to the MVP (*minimum viable product*) advice of launching as fast as possible, but getting feedback can vastly improve the design and usability of your project. You might find an issue that could make or break your project launch.

If getting feedback from others sounds scary, I assure you—the more you do it, the easier it'll be. Go forth and get feedback on what you designed!

CODING YOUR DESIGN

IF THIS BOOK INCLUDED CODE, it would balloon to three times the size! But I did want to talk a bit about some coding philosophy while I have your attention.

Don't worry about being original

A lot of designers complain that too many websites look the same (FIGURE 3-16).

As a beginner designer, *don't worry about being 100 percent original*. Use what works! Sometimes the comforting familiarity could be an advantage, too.

When you grow as a designer, you'll get better at creating new and original details. But when you're starting out, don't fret if your design looks similar to another, like the beautiful (but similar layout) example sites above. No one gets hurt if your design looks similar to something else (unless you've used a design verbatim; don't do that!). Remember, how your design works is more important than how it looks.

Learn the rules and trends before you break them.

FIGURE 3-16: Bootstrap is a popular design framework, meaning that a lot of websites will have similar-looking layouts. This website makes fun of the trend.

Use a CSS framework

Professional designers and frontend developers often scoff at frameworks—most notably Bootstrap (*hellobks.com/hwd/5*), but also Skeleton (*hellobks.com/hwd/7*), Foundation (*hellobks .com/hwd/6*), or PureCSS (*hellobks.com/hwd/9*)—because they constrain your design to fit their plan, often have designs built in (so you're not creating something unique; same complaint as above), and can bloat your code with unnecessary CSS and JavaScript.

But beginner designers benefit from using a framework. It will save you time futzing with CSS, layouts, and best practices. The designed pieces that come with a lot of frameworks (such as Bootstrap) give you a good a place to start from so you don't need to spend a ton of time designing every individual element.

It's more important in the beginning to get a design up and launched as fast as possible. You can always reduce the size of your design code and redesign elements after launch.

Remember responsiveness

We mentioned responsiveness a few times in earlier sections, and it comes into play most when it comes time to code up your site.

Nowadays, with all sorts of devices with all sizes of screens—from tiny phone screens to giant monitors—it's important for your design to work in multiple formats.

Look into media queries (*hellobks.com/hwd/89*) so you can specify CSS rules that only apply to one screen size.

Frameworks usually come with responsive utilities, which is another reason why I recommend them. Pay attention to what features you can use that'll help your design, such as Bootstrap's visible/hidden CSS classes so you can show/hide elements depending on screen size.

Before you launch your site, look at your design in a variety of formats and make sure the flows that you designed work on every size—your design should *work* regardless of the medium (FIGURE 3-17). Coding responsiveness takes a lot of time, but it's important to how well your design works. Don't forget it!

Pay attention to site size

Beautiful images (normal *and* retina), JavaScript, framework code—before you know it, your website can be slow to download. Look at where you can trim fat by reducing the size of your CSS, images, and JavaScript: slow downloading will cause you to lose visitors (FIGURE 3-18).

FIGURE 3-17: Chrome DevTools (*hellobks.com/hwd/90*) allows you to view your design in different sizes in your browser—no need to have the actual device to view compatibility.

FIGURE 3-18: Chrome DevTools also shows you loading times for your website, so you can see what has the potential to slow down the loading of your site.

Use analytics

Don't just launch a design and forget about it—look at how well it's doing after launch. Is your bounce rate (the percentage of visitors who leave your site immediately after viewing) high? Is no one viewing your About page? Critiquing design can be hard because it's very qualitative, relying on gut instincts and personal preference. Adding data through analytics can help you make more quantitative design decisions.

The gold standard in analytics is Google Analytics (*hellobks .com/hwd/48*), but there other analytics packages you can use instead. One option is Segment (*hellobks.com/hwd/49*), which allows you to tie in other analytics services, including Google Analytics and also Mixpanel (*hellobks.com/hwd/91*), and others.

Congratulations on finishing this chapter! I've taken a lot of information and distilled it down into smaller, hopefully more manageable chunks. I hope at this point in the book you feel more confident about making design decisions and planning, prototyping, and building your designs.

4 | CHAPTER 4
REASSURANCES

Dear reader: I have been working on design and websites for around 20 years now and, without fail, this is my mental dialogue on every new project:

> "crap crap crap crap crap crap crap crap crap crap crap crap crap crap crap crap crap crap crap crap crap crap crap crap crap crap crap . . . yes!"

Sometimes it's not a "yes!"—sometimes it's a "maybe?" But the initial "crap" is always, always there.

Design work may never feel easy, especially when you first start out. Your first sketches and mock-ups are more likely to feel terrible than good. Designing, unlike programming, is so qualitative rather than quantitative—we rely on our gut to tell us whether something looks and feels right; you will likely think that your first pieces of work are *not* working right.

Welcome to designing!

This isn't meant to be discouraging—instead, I hope you remember that when you're working on a new design and it doesn't feel like it's going well, you're going through the same process as every other designer. Through iteration, inspiration, research, and work, your design will improve.

When you're thinking these thoughts, you're not a bad designer—**you're *a* designer.**

Don't give up, and keep working. I know you can do it! You'll get better from iteration to iteration, new design work to new design work.

5 | CHAPTER 5
ADDITIONAL RESOURCES

Congratulations, friend! You've made it to the end of *Hello Web Design*!

This was only the beginning. I hope you're excited about learning more about design and using it more confidently in your life. Check out these resources to continue your design education.

Books

The Non-Designer's Design Book, **4th ed., by Robin Williams** (Peachpit Press, 2014): A classic. This book goes into the principles we covered in more detail and covers traditional design concepts that we didn't cover (such as repetition, proximity, and contrast).

Thinking with Type, **2nd ed. Revised and Expanded, by Ellen Lupton** (Princeton Architectural Press, 2010): The definitive guide to typography and type history, covering not just traditional print typography but also web typography.

Don't Make Me Think, Revisited: A Common Sense Approach to Web Usability, 3rd ed., by Steve Krug (New Riders, 2014): Dive into user experience, information design, and usability in this wonderful, easy-to-read book. Provides insights into how users surf and experience the web that will help you build more intuitive and better-working websites.

A Book Apart: *abookapart.com*
Fantastic, short (150 pages or less) books on almost every web topic available—accessibility, frontend development, responsive design, and more.

Blogs and online magazines

Smashing Magazine: *smashingmagazine.com*
Excellent articles and tutorials for web design and frontend development, not to mention a publisher of over 50 eBooks on design.

A List Apart: *alistapart.com*
A bit more on the frontend development spectrum, but *A List Apart* covers a lot of usability, user experience, and accessibility topics that will help you make your designs work better for your users.

User Onboarding: *useronboard.com*
Practical and funny teardowns of onboarding experiences. Great way to see what works (and what doesn't) for major brands and apps.

Online courses

Skillshare: *skillshare.com/browse/design*
This online course website has a wide-ranging selection of design videos and tutorials, from learning Adobe Illustrator to

logo design, lettering, and more. The best online resource for design and visual tutorials.

Theory Sprints by Jarrod Drysdale: *studiofellow.com/theory-sprints*
Great online course to help designers become better designers. If you're looking to jump into a career in design, this is the online course for you.

Inspiration

Dribbble: *dribbble.com*
"Show and tell for designers." Dribbble lets artists and designers showcase their work, with a particular focus on small elements or design details.

Awwwards: *awwwards.com*
Features screenshots of community-chosen web designs—a great place to see beautiful design work and current trends.

Unmatched Style: *unmatchedstyle.com*
Another web design collection site, this one also includes editorial comments on what makes a design great and why it was picked.

UI Patterns: *uipatterns.io*
A collection of patterns that you can use to solve common design and user interface problems, such as designing a date-picker. Most examples are based on mobile interfaces.

Color resources

COLOR PICKERS

Colormind:
colormind.io

Adobe Color CC:
color.adobe.com

Material Design Palette:
materialpalette.com

COLOR THEORY

WebAIM Contrast Checker:
bit.ly/1kVArrR

A Simple Web Developers
Guide to Color:
bit.ly/1RZzK6I

Font resources

TYPEFACE INSPIRATION

Font Pair:
fontpair.co

Beautiful Web Type:
beautifulwebtype.com

Typewolf:
typewolf.com

Typ.io:
typ.io/libraries/google

Canva Font Combinations:
bit.ly/2fsSYA9

TypeSource:
tobiasahlin.com/typesource

WEBFONTS

Google Fonts:
fonts.google.com

Adobe Fonts:
fonts.adobe.com

Brick:
brick.im

Image resources

Unsplash: *unsplash.com*

picjumbo: *picjumbo.com*

IM Free: *imcreator.com/free*

Gratisography: *gratisography.com*

iStock: *istockphoto.com*

Noun Project: *thenounproject.com*

Fiverr: *fiverr.com*

CSS frameworks

Bootstrap: *getbootstrap.com*

Foundation: *foundation.zurb.com*

Skeleton: *getskeleton.com*

PureCSS: *purecss.io*

mini.css: *minicss.org*

Web analytics

Google Analytics: *analytics.google.com*

Segment: *segment.com*

Heap: *heapanalytics.com*

Mixpanel: *mixpanel.com*

Wireframing

UXPin: *uxpin.com*

Balsamiq: *balsamiq.com*

InVision: *invisionapp.com*

GIMP: *gimp.org*

Sketch: *sketchapp.com*

Inkscape: *inkscape.org*

Adobe Products: *adobe.com*

Getting feedback

Five Second Test: *fivesecondtest.com*

Reddit: Design Critiques: *reddit.com/r/design_critiques*

Indie Hackers: *indiehackers.com*

Final thoughts

Please keep in touch with me through the *Hello Web Books* Twitter account (*twitter.com/hellowebbooks*) or through my personal account (*twitter.com/tracymakes*).

More info about *Hello Web Design* can be found on the No Starch Press website: *https://nostarch.com/hello-web-design/*.

Best of luck, and keep in touch!

REFERENCES

For reference, the shortened link URLs throughout the book and their related long URL are listed below.

Chapter 1

http://craigslist.com/

Section 2.1

https://960.gs/
http://getbootstrap.com/css/
http://foundation.zurb.com/
http://getskeleton.com/
http://minicss.org/
http://purecss.io/
https://developer.mozilla.org/en-US/docs/Web/CSS/CSS_Grid_Layout

Section 2.2

http://webaim.org/resources/contrastchecker/
https://color.adobe.com/
https://www.materialpalette.com
http://colormind.io
https://www.smashingmagazine.com/2016/04/web-developer-guide-color/

Section 2.3

https://fonts.google.com/
https://typekit.com/
http://beautifulwebtype.com
https://www.typewolf.com/google-fonts
http://brick.im/
http://fontpair.co/

Section 2.5

*https://www.nngroup.com/reports/how-people-read-web-eyetracking
-evidence/*

Section 2.6

https://www.nngroup.com/articles/how-users-read-on-the-web/
http://www.simpleandusable.com/
https://www.webprofits.com.au/blog/case-study-headline/

Section 2.7

http://analytics.google.com
http://segment.com

Section 2.8

http://www.istockphoto.com
*http://thenewcode.com/944/Responsive-Images-For-Retina-Using-srcset
-and-the-x-Designator*
https://unsplash.com
http://photopin.com
https://thenounproject.com
http://fiverr.com
http://imcreator.com/free
https://picjumbo.com/
https://istockphoto.com
http://www.gratisography.com/
http://upwork.com

Section 3.1

https://www.thebestdesigns.com/
http://unmatchedstyle.com/
https://www.awwwards.com/
https://www.siteinspire.com/
https://abookapart.com/

Section 3.3

https://www.gimp.org/
https://www.sketchapp.com/
https://balsamiq.com/
https://www.uxpin.com/
https://inkscape.org
http://www.adobe.com
https://www.apple.com/keynote
https://products.office.com/en-us/powerpoint

Section 3.4

https://www.reddit.com/r/design_critiques/
http://discuss.bootstrapped.fm/
http://fivesecondtest.com/

Section 3.5

https://varvy.com/mobile/media-queries.html
https://developer.chrome.com/devtools
https://mixpanel.com/

INDEX

960.gs 13

A

A Book Apart 89, 128
Adobe 132
Adobe Color CC 23, 24, 130
Adobe Illustrator 104
Adobe Fonts 35–36
A List Apart 128
analytics 120
Appcanary 79
Awwwards 89, 129

B

Balsamiq 103, 104, 132
Beautiful Web Type 37, 130
Berne Convention 71
Bloomberg Terminal 84
Bootstrap 11, 13, 117, 118, 131
Bootstrapped.fm 114
Brick.im 37, 130

C

call to action 42
Canva Font Combinations 130
centered text 34
Chrome DevTools 119
Citysets 20
clutter 4–5
CMYK 17
coding your design 116–120
Colborne, Giles 56

color 4, 17–27
color psychology 19–20
color theory 17
columns 10
Comovee 80
competitive analysis 64
content 55–62
contrast 21, 82
conversion rates 3
Craigslist 1–2
Creative Commons 71, 75, 76
CSS xvii, 74, 107, 108, 117
CSS Grid 14, 15
CTA. See call to action

D

Daring Fireball 72
design eye 90
display font 30
Don't Make Me Think 128
Dribbble 129
Drysdale, Jarrod 129

F

finding inspiration 87, 88–93
Fiverr 76, 131
Five Second Test 115, 132
Flickr 76
Font Pair 37, 130
Foundation 13, 117, 131
Foursquare 41
"F" pattern 48–49, 53
free font resources 35–36
front-end development xvii

Hello Web Design is set in Avenir, PT Sans, and the Tisa family. The book was printed and bound by the Shenzhen Jinhao Color Printing Co. in Shenzhen, China. The paper is 105gsm matte art paper certified by the Forest Stewardship Council (FSC).